BABEL'S
DAWN

EDMUND BLAIR BOLLES

BABEL'S DAWN

A NATURAL HISTORY
OF THE ORIGINS
OF SPEECH

COUNTERPOINT

BERKELEY

Library of Congress Cataloging-in-Publication Data

Bolles, Edmund Blair, 1942-
  Babel's dawn: a natural history of the origins of speech / Edmund Blair Bolles.
    p. cm.
  Includes bibliographical references.
  ISBN-13: 978-1-58243-641-8
  ISBN-10: (invalid) 1-58243-641-8
  1. Language and languages. 2. Speech. 3. Evolution. 4. Oral communication. I. Title.
  P107.B65 2011
  417'.7—dc22
                                    2011012335

ISBN: 978-1-58243-641-8

Cover design by Ann Weinstock
Interior design by www.meganjonesdesign.com

Printed in the United States of America

COUNTERPOINT
1919 Fifth Street
Berkeley, CA 94710
www.counterpointpress.com

Distributed by Publishers Group West

10 9 8 7 6 5 4 3 2 1

To the people who made the Babel's Dawn blog better than I could:

The many commenters who forced me into a wider perspective and taught me many things I needed to learn about language. And **Laura Newman**, who saved the blog by showing me a way to keep at it without going mad from exhaustion.

CONTENTS

BABEL'S
DAWN

A CONFESSION

WHEN IT COMES to natural history, I prefer museums to books. It's not that I don't love books, but getting from the printed facts of natural history to their breathing truth can often be a bit of a slog, especially when compared with the way a museum sets the visitor down in front of fossils displayed in active positions or dioramas—stages with painted backdrops and three-dimensional figures—depicting lost worlds. Looking at those exhibits, the meaning of the facts pops out at you. So when I began work on my own volume about the natural history of speech origins, the idea of museum displays nestled easily into my head.

Babel's Dawn is organized to help readers pretend they are strolling through a series of museum galleries filled with dioramas that display scenes from the origins of speech. It begins with the last common ancestor we share with chimpanzees (from about six million years ago) and proceeds on down to the first storytellers (a bit more than a hundred and fifty thousand years ago). Naturally, I have imagined a modern museum, and when visitors arrive in the entrance hall they are handed devices called audio guides, complete with headphones. Besides pretending that you are looking at scenes in dioramas, pretend that you are listening to an audio guide that provides the facts and ideas connecting the displays. It is that simple.

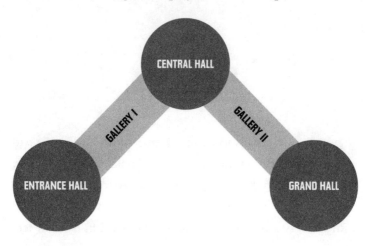

So let's tour my natural history museum's exhibition on the origins of speech . . .

ENTRANCE HALL: BECOMING DIFFERENT

L IKE ALL SCIENCE stories, this one begins with wonder. Over the centuries, many people have noticed an infinite chain of language acquisition. We learned to talk from our parents, who learned from their parents, who learned from their parents—but how did that chain begin? Asked in modern, Darwinian terms, we know that our ape ancestors did not speak, so there must have been a time when members of the human lineage began speaking without learning it from their parents. How did they do that?

Against a wall stands a wax figure of Samuel Johnson, famous as the author of the first great dictionary of the English language. He wears an eighteenth-century wig and a cheap frock coat. Behind him, written on the wall, is a quotation from one of his famous conversations.

Johnson, like many people, tried to explain the birth of language as a miracle, proposing that language began as a product of divine inspiration. The wall quotation reads:

A thousand, nay, a million children could not invent a language. While the organs are pliable, there is not understanding enough to form a language; by the time that there is understanding enough, the organs are become stiff. We know that after a certain age we cannot learn to pronounce a new language.

The argument makes sense. Children don't have the brains to come up with language; adults don't have the tongues. But it turns out that Johnson underestimated how creative groups of children can be.

Most languages spring from the abyss of time, the way French and Spanish stem from Latin, which descended from an old Italic language that was one of many offspring of the Indo-European tongue that had ancestors of its own. But there are a group of languages known as Creoles that are much more recent. They were spawned by the children of slaves and indentured servants.

Slaves may have been perfectly articulate when they were captured, but conditions changed. An African Ibo tribesman, for example, might have been dragged to a plantation in the West Indies where he was worked by people who spoke an incomprehensible white-man's gibberish. The tribesman was surrounded by other slaves from different parts of West Africa who knew nothing of the Ibo language. He was forced to speak a pidgin—a hodgepodge of words without grammatical associations. The obvious solution was to come up with grammatical usages that let people express more complex ideas and relationships than pidgins can organize. A grammatical pidgin is called a Creole language, and they have emerged wherever communities were once forced to communicate only through pidgins. About thirty years ago, it was finally established that Creoles were created by the children of those pidgin-speaking adults.

A statue of a girl who looks about eight years old shows her making the sign for "water" in Nicaraguan Sign Language.

About twenty years ago, a linguist named Judy Kegl (now Judy Shepard-Kegl) happened to be on hand in a Nicaraguan school for deaf children where she was surprised to observe the children create a sign language. A centralized school for teaching deaf children was new to the country. Previously, deaf children tended to lead isolated existences, getting by with a few signs known as home signs, a kind of pidgin for deaf people. The children were brought together in the hope of teaching them proper Spanish, but instead they turned their home signs into a full-blown functioning sign language.

Dictionary Johnson was wrong. When grouped together into a community, human children have it within themselves to take random words from their environment and create a new language. That's how Creole languages currently spoken in the Caribbean, South America, Hawaii, the Indian Ocean, and Africa began. It is decisive evidence that no miracle is required to produce a new language, and it suggests very strongly that there is some inborn tendency to express ourselves. As Kegl put it, we have a natural "hunger for language." In other words, language has a biological (and therefore an evolutionary) side to it.

Side by side are statues of Charles Darwin and Alfred Wallace, the co-discoverers of the theory of evolution through natural selection. Wallace hands Darwin an envelope, symbolizing the historical moment when Wallace mailed Darwin a paper setting forth his theory. After years of silence, Darwin was forced to report his ideas publicly.

Darwin had suspected that language was biological, but most people thought that languages were entirely cultural. As late as the 1950s, a commonplace of linguistics held that the fundamental fact of languages was how different they all were. Whenever a new language was discovered, experts assumed that it might be completely unlike any language previously known. They took it for granted that language had begun as a result of invention, a prehistoric event whose birth was lost to discovery and not to be inquired about.

The doctrine of language's purely cultural nature and origins gave way because of theoretical work by Noam Chomsky, the dominant linguist of the last half of the twentieth century, and because of the fieldwork in Creole languages by another linguist, Derek Bickerton. In his zesty book *Bastard Tongues*, Bickerton tells how he traveled through the world studying Creoles and their histories until he had the proof that different Creoles had different origins. They are not simply varieties of one common language. He also showed that it was the children who transformed them from pidgins.

How do children accomplish such a feat? Even more fundamentally, how do ordinary children start speaking so effortlessly? No account of language origins is likely to be accepted unless it explains why it is that (with a few tragic exceptions) every human baby in the world starts using language without requiring special training. Meanwhile, no animal, even with extensive drilling, manages to use, at the absolute best, more than a few hundred words.

The most obvious answer to these basic questions is that humans have some kind of instinct for language. By the early 1990s, there was a widespread view that language was entirely instinctive and that no two languages really differed all that much. Chomsky still holds that opinion, but it has become a minority position among researchers into language evolution.

One much disputed point concerns the amount of biological change necessary to produce modern speech. The language-instinct proponents in particular have often favored a single genetic mutation that, in a sudden "big bang," transformed our ancestors into symbol users. Others, particularly biologists, criticize that view as naive. There is much more to talking than getting the grammar straight, as this tour will show. We had to alter our windpipes, take subtle control of our tongues and lips, tune our ears to vocal sounds, alter our breathing patterns so we could speak for extended periods without becoming giddy, become willing to listen to what was on another's mind, become willing to tell others what we knew, develop brains able to use thousands of words, master paying joint attention, and obtain a knowledge of grammar. And it was all done by the normal, ceaseless process of changing generations—perhaps almost 475 thousand generations between today and our last common ancestor with chimpanzees.

A picture on the opposite wall shows a chimpanzee dressed in magician's costume pulling Shakespeare (whose head is visible) from a hat.

The wonder of evolution is that every link along the chain of generations is almost exactly like the one before it and the one after it, yet the start and end of the chain can be very different. In the case of language, we began with an ape and ended with us. All along the way there was a steady mixing of at least three generations—juveniles, parents, and grandparents—and

the juveniles always looked pretty much like the grandparents.
It was a very slow morphing through a long line of individuals.

There are two natural—and misleading—ways to tell the
story of evolutionary change. One focuses on the abstract turning points and finds revolution. For speech origins, *revolution*
means a gateway through which the human lineage passed.
On one side were the non-talkers, on the other blabbermouths.
Revolutionary speculations are full of references to hopeful
monsters and big bangs (or, for the technically minded, *salta-
tions*), events that made a knife cut across history and sliced the
connection between what came before from what came after. To
justify their position, revolutionists point to the many obvious
differences between apes and us, including our success at adapting to many eco-niches, our languages, and our moral codes.

Their rivals tell a slow story of continuity. Sure, we talk,
but all animals communicate. True, we make machines, but
chimpanzees and many other animals use tools as well. Yes, we
love, but bonobos give one another reassuring hugs. Stories of
continuity put the stress on modification and similarity. They
point to logic to justify their claim. You cannot evolve something from nothing, so obviously we must be the result of what
came before—tweaked and stretched, but nothing really new.

Both arguments make sense, and both are wrong for the
same reason. They overlook evolution's essential mechanism
of change: variation combined with selection.

Variation without selection occurs all the time. Variation
happens because the systems for biological reproduction are
not exact. The windpipe between mouth and lungs, for example, will differ slightly in every *Homo sapiens*. Usually
the differences average out and the typical windpipe remains

stable over time. Occasionally, just by chance, the variations lean more one way than another and the shape of the average body's breathing organs drift a little. The important truth about variation, with or without drift, is that it is random and tells us nothing about the larger world.

That larger world has its say when it comes to selection. *Selection*, by the way, is a poor name for what happens. It sounds like somebody or something actively makes a choice, but in fact nothing and nobody is in charge. The process is much like a sporting playoff in which competitors face off and one team ends up as champion. Nobody selected the champion; it just won. *Natural success* might be a more precise phrase, but at this late date in history Darwin's term is here to stay.

Selection begins with variation. A variation in a windpipe might be rejected for millions of years, and then suddenly circumstances change and the variation is passed on to another generation as just the right thing. What makes a variation the right thing? Something about the environment must have changed, making what was once an unwelcome trait become a winner. Selection keeps generations of organisms in tune with the environment, even as the environment changes. Selection works like a cruise ship's stabilizers, making constant, small adjustments so quickly in response to irregularities that the passengers barely notice them. Some people, however, think of evolution as a ship going somewhere and assume that it can make progress. Others think of evolution as getting nowhere and assume that nothing can change.

No more prestigious example of progress-oriented thinking can be found than in Noam Chomsky's own speculation about language origins. His ideas are unusual because he proposes that language began as a medium for internal thinking rather than for communication. He said in a talk a few years ago:

> The simplest account of the "Great Leap Forward" in the evolution of humans would be that the brain was rewired, perhaps by some slight mutation . . . There are speculations about the evolution of language that postulate a far more complex process: first some mutation that permits two-unit expressions, perhaps yielding selectional advantage by reducing memory load for lexical items; then further mutations to permit larger ones; and finally the Great Leap . . . Perhaps the earlier steps really took place, though there is no empirical or serious conceptual argument for the belief. A more parsimonious speculation is that they did not, and that the Great Leap was effectively instantaneous, in a single individual, who was instantly endowed with intellectual capacities far superior to those of others, transmitted to offspring and coming to predominate.

There are many things in this passage likely to trouble anyone familiar with the process of evolution. The joking allusion to the Maoist term "Great Leap Forward" leaves no doubt that this line of thinking is in the revolutionist, evolution-as-progress tradition. On a more technical level, there is all that emphasis on variation rather than selection. The engine of Chomsky's revolution was mutation. Suddenly we get a variety of individual "who was instantly endowed with intellectual capacities far superior to those of others." (Remind you of anybody?) But the issue of selection is completely missing. Yet we cannot just assume that the superman wins simply by the nature of his distinctiveness. Remember the warning from the preacher in Ecclesiastes 9:11:

... the race is not to the swift, nor the battle to the strong, neither yet bread to the wise, nor yet riches to men of understanding, nor yet favour to men of skill; but time and chance happeneth to them all.

Without selection, time and chance doom the brightest hopes. Eventually the race may go to the swift, but only when there are enough predators to catch the slowpokes. Without selection, time and chance rule the hour.

For us today, the benefits of language are so obvious that it is tempting to downplay selection. We assume that any variation that supports speech, even a little bit, will be favored, but remember that while we talk plenty, other species do not even talk a little. There must be very powerful reasons why time and chance have never brought speech to any other species. Surely lions would be better hunters if they could talk among themselves and develop more promising plans, but they do not speak. Even if Chomsky was right about how the mutation worked, there must have been some reason why a bit of brain rewiring was selected this time around.

The continuitarians can laugh at the revolutionists and their misreading of how evolution works, but they have made a great mistake themselves. Their motto comes from a French wit who said in response to a political upheaval, "The more it changes, the more it's the same thing." It was a sardonic cry of despair, picked up by conservatives who like to deny that revolutions change anything, and it carries proponents right back to Ecclesiastes: "Time and chance happeneth to them all." What they miss is the power of selection to effect great changes over time.

The preacher in Ecclesiastes assumed that green plants, swimming fish, and bird migrations had been there from the beginning, so he could insist that there was nothing new under the sun. We, however, suppose that the world began as a hot, sterile mass orbiting a new sun, so the presence of plants, fish, and birds argues that time and chance can accomplish wonders when you mix selection into the pot.

The most extensive argument for the slow-and-steady position was made by Christine Kenneally in her book *The First Word*. Unlike Chomsky, who takes language to be a coherent system of syntactical rules, Kenneally sees speech as comprising a "suite" of physical abilities, each of which we have in common with many other animals. She writes:

> In most disciplines the focus used to be on the separateness of animals and humans, that gulf being marked most strikingly by language. But over the last few decades, the emphasis has switched to investigating the continuity of life *in addition* to clarifying the boundaries that lie between species. We no longer have a sense that we are standing apart from all animal life and that language is a discrete, singular ability that isolates us . . . [Many animal researchers] talk in terms of a rough continuum between modern animals and modern humans, describing the differences between them and us as more quantitative than qualitative. Such a continuum . . . is based on the existence of similarities and differences of features important to language.

And Kenneally is quite good at presenting the many things language requires that are found in other animals as well, but she is unlikely to persuade a reader who starts with the idea that there is a qualitative difference between ape and human communication. A list of similarities between the two is apt to be dismissed as noting only secondary features of language, none of which get to the core of what makes human speech unique.

It is in explaining why humans and only humans talk that the inadequacy of both revolutionary and continuitarian accounts become evident. Technically speaking, they are both forced to argue in a circle: Humans alone speak because no other animals do. In the revolutionary case, humans alone speak because no other animal got the appropriate mutation. Why didn't they get that mutation? They just didn't. Meanwhile, the continuitarians say humans alone speak because no other animal has taken the requisite set of skills as far as we have. And why haven't they done so? They just haven't.

More generally, these circular arguments miss what is the central point of this tour through the natural history of speech. Humans alone speak because we alone need to speak. Language supports something essential in us that is not even trivially necessary for the other species of the world. The revolutionaries were right in arguing that there was a break. Something new appeared under the sun, but they became trapped in their explanatory circle because they did not try to find what had changed that not only made language possible but necessary. If it were not for language, our lineage would have died out on the African plains. Indeed, the other erect, bipedal apes on the grasslands did become extinct. Yet non-talking nonhumans have prospered. Why don't they need language too? We know from specialized training projects with captive apes that they can use sign language about as well as a two-year-old, yet in the wild they never use gestures to form words. Nor do chimpanzees who have been taught to sign chatter among themselves. That makes for a Tom-Dick-and-Harry trilogy of mysteries.

Tom: *How did people come to talk?*
Dick: *How can children use language so effortlessly?*
Harry: *Why don't any other animals speak at all?*

It is the Harry mystery that trips up the continuitarians. They were right when they said that we are an extension of our primate ancestors, but they failed to notice that the difference between our ancestors and us goes far beyond speech. Speech is the essential instrument for holding human communities together, but there is more to being human than using language.

A wall displays a triangle whose corners are labeled "speaker," "listener," and "topic." Arrows point between speaker and listener, and separate arrows point from speaker and listener to topic.

The speech triangle summarizes the community structure that distinguishes human communication. Machines and animals communicate to manipulate one another; two biologists, Richard Dawkins and John Krebs, have even written that animal communication is so manipulative and controlling, so unlike human communication, that they are "tempted to abandon the word *communication* altogether." Humans communicate to pilot attention to some topic of joint interest. Apes have a two-way communication structure. To get another's attention, a chimpanzee might slap the ground. Other apes naturally look toward the unexpected noise, and the first ape can then make a begging gesture or give a look of intimidation. This behavior is typical of procedural communication. The first step, the slap, is followed by a second step, and both steps are focused on manipulating the other.

Humans can beg and intimidate as well, but they can also do things that apes do not do, by using the speech triangle. Its three-sided structure supports a communicative function unknown elsewhere in the animal world: mutual consideration of a topic. When we think with words, we think about something: the topic. When we converse with others, we have a topic in mind even if it keeps changing as the conversation rambles on. The kind of communal knowledge created through the exploration of a topic is the fundamental gift of language. The formal study of language usually concentrates on the abstract elements (notably syntax and symbols) that organize sentences, so linguists rarely worry about how these elements originate. But a look into speech origins quickly shows that the keystone supporting the whole triangle is our ability to join with one another in considering a topic.

Psychologist/anthropologist Michael Tomasello describes an example of the difference between human helpfulness and the rugged individualism of apes:

> ... when a whimpering chimpanzee child is searching for her mother, it is almost certain that all the other chimpanzees in the immediate area know this. But if some nearby female knows where the mother is, she will not tell the searching child, even though she is perfectly capable of extending her arm in a kind of pointing gesture. She will not tell the child because her communicative motives simply do not include informing others of things helpfully.

Why aren't chimpanzees motivated to help? There is a straightforward, Darwinian explanation for the ape's *mum's-the-word* behavior. Individuals don't help non-kin. There is nothing in it, no survival or reproductive advantage, for the informed adults to help the whimpering child of another. And yet humans typically do help out whimpering children, even if the child is a stranger. An adult, happening upon a solitary, unknown, whimpering child, is very likely to stop and ask what is wrong, take charge, and stay with the child until the problem is resolved. This activity strikes us as perfectly natural, normal behavior, even though it is contrary to so many practices of other animals.

No, this tour is not about to challenge the theory of evolution, but it does say we need a good evolutionary account of how a species of ape with no motive to assist others outside the family became a species that takes such group helping for granted. How is it that our old behavior—ape behavior—seems shocking? Six million years ago, apes already had the physical ability and the brains to offer some help to one another, yet they did not help. The first part of the rise of speech, therefore, was the evolution of the speech triangle and a willingness to share what you perceive.

GALLERY I

Sara | 6,000,000 Years Ago | Generation 0

An ape with brown-and-white fur and a crumpled left ear sits in a tree while a still-nursing child hangs on to her. In the same tree, not too far away, stands a second offspring, perhaps eight or nine years old, tending to its own affairs.

Call the nursing ape Sara, although she lived long before anything had a name or the need for one. She was a forest ape of some uncertain species and the mother of at least two young who survived to become parents themselves. The brown-and-white fur in the diorama is a pure invention, but so would any other color be. The commonplace choice would be to make her black-haired, in keeping with gorillas and chimpanzees, but Sara was neither a gorilla nor a chimpanzee. Giving her a different color stresses the point that she is her own distinct species. Until recently, Sara—or to use her technical ID, the *Last Common Ancestor* of chimpanzees, humans, and bonobos (often called the LCA)—was usually thought to be very like a chimpanzee; however, recent fossil and chromosome evidence suggest that chimpanzees may be as unlike Sara as we are.

The crumpled ear is included because, like all individuals, Sara was physically a bit different from her contemporaries.

The ear is a stand-in for whatever physical distinctiveness the real Sara had. All concrete details in these dioramas serve as metaphors for something that can only be known abstractly and generally, but the truth of natural history is not abstract. There was a real individual who lived and produced at least two surviving offspring. One of those survivors became the founder of the human lineage; the other started the chimpanzee/bonobo lineage. Stated so theoretically and dispassionately, we risk missing the central point, which is that despite its having been so long ago that it sounds like somebody else's dream, Sara really lived.

How long ago was Sara? All dates in this story are approximations, with Sara's date of six million years ago among the least certain of them all. The number is based on DNA evidence, but studies of the ape-human divergence have used DNA evidence for over forty years and found a range from eleven million to four million years ago. The exhibit's six-million-years-ago figure is based on eight studies of nuclear DNA conducted in the twenty-first century. Their dates range from 6.8 million years to 5.4 million years ago, with a median point at 5.9 million years ago. A study published in late 2010 put the date at eight million years ago. The authors combined the uncertainty of DNA dates with the uncertainty of fossil descent and arrived at a date from a probabilistic model. It is one more addition to a growing number of voices. In this tour we will call it six million and move on.

How much like us was this animal? As the last common ancestor of both humanity and the chimpanzees, Sara holds an almost mythic position in our imaginations, or would if she were not an ape. How much difference does that family membership make? According to a study published in 2002, chimpanzees share 95 percent of our DNA. Some other studies put it a few points higher. Is that a lot or not? It sounds like a close match, but what are the standards for making sense of this technical information? After all, mice match us 85 percent, and the duckbilled platypus shares 82 percent of our genes. We do not yet know how to interpret such information. We once thought that DNA was like a blueprint, but that was wrong. None of the other analogies—e.g., recipe, set of instructions—seem quite right either. We still know so little about the translation of DNA into an organism that we do better by comparing the physical results rather than the code.

> Sara stands out on a branch. She holds another limb with her right hand while tugging on a vine with her left. The vine brings a higher branch down, carrying its fruit within grasping range. Sara's baby clings to its mother and pays no attention to her actions, but another juvenile watches Sara's motion attentively.

We sometimes overstress the need for speech, which risks giving the impression that it is the only way groups can pass on their accumulated wisdom. But long before there was speech, apes and monkeys benefitted from social learning. Such learning appears to depend on the ability of an inexperienced ape to take notice of something like another's use of a vine and recognizing the action's utility. The younger ape does not say, "Ah, so!", but it remembers and learns.

This kind of learning is a step beyond the basic learning of associations built on a reward that comes from a behavioral response to an environmental stimulus. A classic example of that kind of learning is a pigeon in a laboratory who discovers that by pecking a red button it can receive some food. The food is the reward and is associated with the red button and pecking. In social learning, the monkey imagines the reward and sees the association. Perception is enough; actual physical experience is not required.

Sometimes this kind of social learning in apes is called *culture* because it survives for generations without being passed along genetically. Others have a more complex definition of *culture* and disagree, but already by Sara's time the founders of the human lineage were not limited to doing what was built into their DNA. Their learning could be based on just paying attention.

> With her infant clinging to her chest fur, Sara bends forward and puts her face almost directly into her nine-year-old's face. As she does this she makes a heavy-breathing sound.

Sara controls her young, as much as they can be controlled, through sounds, gestures, and actions. Confronted with this look, the juvenile is likely to follow Sara obediently as she moves away. The sound made by the mother seems insignificant when compared to the force of that stare, but Simone Pika, an experienced student of ape gestures, reports that vocalizations are a common part of voluntary gestures from apes. The sounds in these cases seem to be offered as attention-getters, the equivalent of saying "Ahem" and holding out one's hand. Sound and gesture together is much older than humanity.

The mother-child relationship in the display may seem familiar to humans, but there are surprises. Sometimes a juvenile will beg for food from its mother, holding out a hand toward her. Sara might give something, but she is not generous by nature and the food she offers is often a very meager portion. Her brute selfishness can be shocking, and there are even times when she reaches into her child's mouth to pull out a half-chewed wad for her own consumption. At the same time, the males, who are considerably larger than the females, often try to rob Sara by chasing her away from the choice food, or from an admirable perch if, on some whim, one of them wants the spot.

With no reason to let others know what she knows, the ape that parented both the chimpanzee and human lineages was left with very little to communicate outside of her own needs: *Give me food. Scratch me here.* Sara quite likely made gestures for those sorts of messages, but communication about an outside topic like gossip, sports, plans, arts, ideas, travels, and on and on—no, there was none of that.

Ape communications tend to be unequal, for the producer is trying to control the receiver. Thus, the most common communications among the apes reflect unequal relationships: mother to young, dominant to subordinate, or supplicant to advantage-holder. The kind of peer-to-peer exchange that dominates speech is almost unknown among apes, with the possible exception of warnings about dangers. Even during grooming, when apes go through one another's fur to remove any insects, to clean sores, and to spruce up the fur, there is nothing like the exchange of pleasantries found in barber shops and hair salons around the world.

Speech works differently than the communication of other animals, even if the animal communicates with sounds. A bird singing, a gibbon calling, and a lion roaring are all communicating, but they are not communicating in the intimate way humans do. They are, so to speak, broadcasting their message. A birdsong signals to anyone who might hear, *This space is mine.* Human speech holds speaker and listeners together by focusing each other's attention on something real or imaginary.

"Sorry," a person might say, "I was not paying attention." Human communication can fail at this most elementary level. Nobody ever says anything similar to a roaring lion.

> Sara sits quietly grooming her older offspring while from nearby she hears the quarrelsome cries of two unseen males who sound as though they are facing off, although not yet fighting physically.

One of the surprises of ape studies has been the simplicity of their vocalizations. Compared with some less intelligent monkeys, apes don't say much. Vervet monkeys in East Africa have smaller brains, but make a range of signals that distinguish between threatening predators. It has also been shown that vervets take context into account when they respond to an alarm. South American monkeys, such as the cotton top tamarin, tack sounds to the front or end of a signal to alter its meaning. Yet the much-smarter apes signal much less. Why is that? The answer may have been found by primatologists Robert Seyfarth and Dorothy Cheney, who note a dilemma that the apes face: Just as a trembling voice can reveal more than the speaker intended, every ape vocalizer risks just such exposure. Ever since Darwin, people have assumed that ape knowledge is quite limited because their calls are reflexive and emotional. However, the fact that the source of the information is emotionally triggered does not mean that the listener cannot draw logical inferences about the caller and the social situation. The vocalizer is sharply limited in its ability to produce calls, but the ape listener is not anywhere nearly so strongly constrained in its ability to understand the richness of the situation producing the calls. Apes are very capable at perceiving, navigating, and drawing logical conclusions about their society and environment. Some scholars have argued that humans, with three times the brainpower of a chimpanzee, must inevitably speak, but the pattern among other primates suggests that the smarter the species, the more careful it has to be about tipping its hand through overly informative vocalizations. If that pattern had held, perhaps adult humans would be limited to cruel laughter, desperate weeping, and screams of agony.

> Sara sits out on a branch picking figs to eat. She is skillfully attentive as she works, taking the fattest fruit first, in case some male suddenly comes beside her and reaches for the juiciest remaining figs. There is an alert deftness to her motions. She holds on hard enough to ensure the steadiness of her grasp, but not so hard that she squeezes out any of the pulp. If a fruit stem resists her pull, she strengthens her action just enough to keep the motion smooth and steady. To an observer, the whole operation can look mechanical, but that is an illusion brought on by Sara's steady concentration on her task. She does not eat like a gourmet who focuses on the subtleties of taste, but she is aware of the pleasures of the fruit and alert to any misflavor that might warn of something wrong. Meanwhile, her youngest holds on to her, oblivious to the danger of their high perch. Sara can feel the warmth of her youngest nestled against her own furry bulk. Around them, among the branches of ripening figs, sit half a score of other troop members helping themselves to the tree's bounty.

The deliberate attention Sara displays is a very old power of mammals. It is not only apes, or even primates, who can be so deliberative. Predators scan a herd of prey, judging what they see, looking for a vulnerability. A beaver building a dam pays attention to how a branch fits into the whole. It is this capacity for deliberate attention that shows that an animal can be fully engaged in the present moment. An elephant reaching with its trunk high into a tree, stretching and lifting its whole body to grasp and snap a branch, reveals the way an animal can adapt its actions to the details of the setting. Sara eating figs shows the same thing. Humans are the grandmasters of deliberate attention. A skilled woodworker, for example, can spend many hours creating and perfecting a comfortable chair so that the joints fit exactly, the back arcs properly, and the arms are gracefully decorated. This rich, productive attention span is a gift we have received from our long evolutionary history.

> Sara stands on a branch while facing away from the tree trunk. A python is coming down the trunk. Something about the snake must have alerted Sara because she has turned her head to look straight at it.

Again we see Sara paying attention to something, but this time she does not use deliberate attention. Nothing about her behavior is voluntary. Sara sensed something—perhaps a sound, perhaps an odor—and reflexively directed her attention to its source. Reflexive attention is handled by a different part of the brain and is probably older than deliberate attention. It seems to be in every mammal's repertoire. A sudden sound will make a grazing animal pause and look toward the noise. That power can fool a person into supposing that attention is passive, simply waiting for information, but as the Italian philosopher Giorgio Marchetti insists, attention "is a kind of activity." Once brought to attention, an alerted animal works at deciding what to do. You can see the difference between attentional capabilities and reflexive powers in the legendary duals between cobra and mongoose. A snake responds immediately to its senses. Meanwhile, once the mongoose comes to attention it takes its time, looking for opportunity. Most of the time it is the alert mongoose who kills the lightning-fast cobra.

A special kind of attention—joint attention—combines reflexive and deliberate attention, and joint attention is crucial to language. It is not simply the mass attention of a group of animals or people looking at the same thing. When zebras drink at a water hole and hear a sound, they all jerk their heads up in reflexive attention. Such behavior in a group is mass attention. Joint attention comes when people know they are paying attention to the same thing and deliberately inter-*re*act. For example, two campers at a water hole will, like zebras, react reflexively to an unexpected noise, jerking their heads and listening. Then mass attention becomes joint attention as the two look toward each another. Now they are consciously attending together as members of a group and looking to each another for cues on how to behave.

Chimpanzees in captivity can also engage in joint attention with their keepers. They can learn to point at things and name things using hand signs. So they are social enough and have the smarts to use joint attention, yet observations of free chimpanzees in the wild show that they almost never use it. Humans, on the other hand, depend on joint attention for survival. We use it as a tool. For example, in a group formed by

Robert, Ben, and Martha, Robert points out something. Ben adds a comment. Robert points to a new detail. Martha adds her observation. And suddenly Robert, Ben, and Martha all share a new plan, thanks to their joint attention.

If chimpanzees and humans are both capable of using joint attention, we can assume that our last common ancestor, Sara, could also engage in it. So the question is not how we evolved joint attention, but how we became efficient at using it and dependent on it.

A psychologist at the University of Miami, Peter Mundy, and one of his graduate students, Lisa Newell, have done groundbreaking work in learning how the human brain supports joint attention. They recognize the two kinds of attention we have seen in Sara: reflexive and deliberate. Non-specialists call both the reflexive and deliberate systems by the same name, *attention*, but they evolved separately, they develop separately during a creature's growth, and, excepting for ourselves, they operate separately.

During speech, these two distinct attention systems function as a unit. A speaker utters a string of words that directs a listener's reflexive attention. "What about that six ball?" one pool player asks an instructor. The listener's eye shifts automatically to the ball, just as it would if the ball had suddenly moved, but the examination of the ball is not limited to the brute survival questions of reflexive attention. The more subtle powers of deliberate attention can be brought to bear on the object because of neurons that work like a cable linking the two attention systems.

The instructor becomes the speaker and says, "By kissing the five ball there [and the listener's eyes dart to the five ball], the six ball should fall into the side pocket." And the original speaker, now the listener, is able to visualize what he hears. *Ah.* A lesson in how to do something is passed on verbally. Speech raises the social learning of the ape to a skill beyond the limits of the passing moment. It was made possible by taking advantage of the wiring that has integrated two separate parts of the brain.

And there is how evolution can produce a distinctly different creature without having to overhaul the entire DNA molecule. Our brain's two attention systems are substantially the same as what Sara had six million years ago. What is new is the circuitry that links them and turns two separate abilities into a unified function that can explore the world in a new way. People sometimes fear that if they really understood how the brain worked and how we differ from apes, the answers would be disappointing, like learning how a magic trick works. But magic tricks are ruined because the explanation shows us that there is no magic, just plain physics with some tomfoolery thrown in. The brain, however, is no magic trick. We really do speak, understand, and learn about things beyond the brute facts of survival. So, instead of exposing a trick, we can sharpen our appreciation of just what it is that makes humans so unusual a species.

Our capacity for directing one another's attention makes joint attention worth having. We learn from one another's perceptions, emotions, and thoughts. If you can't learn from the group, there is not much point in looking toward other members of a group when one is startled. But if a new brain circuit was the only thing needed for joint attention, it should be much more widespread than it is. Every social animal species should have evolved such a circuit. The fact that they did not implies that more than brainpower is at work.

> Sara looks on from behind as four large males shake branches; in another tree, four apes from another troop scream back. The air sacs around the apes' larynxes throb like bagpipes and draw the support of any males close enough to come running. Sara is used to seeing one or two late ones hasten to the scene and join in the uproar.

All this screaming might seem like a forerunner of language, but it is not. The cries are not cultural, but biological and automatic. They trigger a reflexive behavior—in this case, more screaming by other apes. There is no speech triangle, no joint attention, no common topic. And the response of an ape to the cry is quite predictable: More apes will join in the howling. It is serious business, but not language.

In many ways the males in Sara's troop are pests who use their size to take what they want, but they have their benefits when facing enemies. Sometimes, if a section of the forest in Sara's territory is ripe with fruit while a neighboring territory remains barren, a rival troop will look for ways to steal food, but Sara's troop remains alert for invaders. By Sara's time the forest had been shrinking, and some territories that were once able to support troops could no longer do so. That change put great strains on the rival troops as they desperately sought ways to eat. Boundary clashes along the edge became more frequent and bitter. A great extinction had begun. Sara, however, behind her wall of boundary guards, endures, and her children will grow to have children of their own.

Apes had thrived in African and Asian forests for about fifteen million years. That is almost three times the span covered by this tour, and almost six times the span of the *Homo* genus. The age of apes ended because the range of the African forest began to retreat. Even then their descendants persisted. One line stayed in the remaining forest; the other spread over the world.

Ape success comes from the way their societies are organized around the brute facts of biological existence—facts like death, hunger, sex, and space. Apes do an excellent job of reducing the outside threats by keeping a weather eye out for predators, bringing many different observers to the task of finding food and fending off invasions of their territory from neighboring ape troops. What ape societies do not mitigate are their internal facts—the competition for sex, food, and status. The politics within ape groups makes it pointless to share the kind of information that is common among humans. Their societies are united against the outside world, but contain ferocious individualists when it comes to each member's own interests. They may forge coalitions, but these alliances are subject to cessation or betrayal at any moment. Theirs is the logic of Lord Palmerston: They have neither permanent friends nor permanent enemies, only permanent interests.

This ferocious individualism extends to the operation of evolution itself, and explains why the shared capacities of joint attention never evolved in other primates. It is not that apes never evolved anything that helps the group. For example, they make an automatic cry when they see some food. (We know the cry is automatic because chimpanzees have been observed trying to muffle their call, like a person trying to hide a public burp.) But there are no penalties for putting oneself above the group. Apes do not shun or push away another ape who has been caught trying to keep all the food for itself, nor is there a penalty for standing back and not participating in the defense of the troop's border. Surely it would be a disaster for the group if too many males took a let-Joe-do-it attitude, but there is no evident pressure from the group to help out.

Such an absence of group pressure is notably different from human communities where demands for mutual respect and cooperation play such an important role. The price of speech, for example, is having to listen to the concerns of others and even having to take the concerns of others into account when speaking oneself. The give-and-take of mutual listening and courtesy may strike most humans as far too minor and natural to be called a sacrifice, until we look at other primates and realize they never do it. They never turn their attention away from the brute facts of their individual existence.

> Crumpled-ear Sara sits out on a limb. Her older juvenile is gone from view. A recently born infant sucks at her teat. On a branch above her sits a small dark bird, but she pays it no attention. In a moment it may fly off in a flutter of wings, but Sara has seen birds before and is unlikely to watch it go. The boundary tensions have not ceased. Somewhere beyond her space, trees are giving way to more open country. A long drought has been killing whole sections of the seemingly eternal forest. Sara has never known what it is like to see two hundred yards stretch ahead of her, but on the other side of her troop's border are apes who have seen such a sight, and they create a steady pressure to move further into the forest. Even in ordinary times the territorial boundaries of ape troops are not fixed, and Sara's territory does slowly change its shape. Sara, fully grown, prefers the areas she has long been accustomed to, and she is content helping herself to figs and

termites in the time-honored way, using her well-honed skills and staying within the world that is always present.

Harun | 5,999,928 Years Ago | Generation 6

An ape, substantially larger and more robust than Sara, because he is a male, but of the same brown-and-white species and with a crumpled ear like hers, stares toward the base of a thorny bush where a warthog is digging a hole. The bush's thorns are long daggers, not small, rose-bush-sized hooks. They work like porcupine quills that point in all directions and cry out, *Stand back!* In the pit where Harun looks, part of a fat tuber is visible.

Although this scene does not look like a move toward speaking, evolution lacks a destination and does not go in straight lines. The important fact here is that serious change of some kind has come to our story. Natural selection is typically a conservative process that rejects almost every change to the genome of a successful species. A successful species has to balance many thousands of features and behaviors. Changing one can disrupt many others, so the one thing does not change. We can see this equilibrium in our own lives. For example, drivers in some countries keep to the left, and in other countries they keep right. If all countries followed the same rule, it would be easier for travelers and car manufacturers alike, but changing would disrupt so many habits, road signs, freeway exits, and existing automobiles that the change is virtually impossible. In biology, this kind of resistance to change is known as an evolutionary stable strategy, and once one is established it tends to persist. Before natural selection lets major innovations disrupt the equilibrium, something has to happen to turn a successful species into a struggling one.

Sara was successful, giving birth to numerous young who survived and were fruitful themselves. But only seventy years later, Sara's descendants along the human lineage are struggling. (All of the individuals named in this tour are, like us, Sara's direct descendants.) This woodland is not the same as the brush country that modern travelers in Africa often drive through on their way to open grassland. For one thing, it is wetter and not especially dusty. The modern Sahara, which is the source of much dust, was still millions of years in the future. Trying to survive in eastern Africa was not as difficult as eking a living from today's semiarid, dusty, thornbush land, but it was radically different from living in Sara's forest.

The earth had been cooling for ten million years before Harun's generation. The change was part of a long process that would climax only millions of years after Harun with the onset of a series of ice ages. In Africa the change in climate eventually transformed forests into open woodland, where Sara's great-great-grandson was forced to adapt.

Sara spent most of her life up in the trees eating ripe fruit. Harun lives mainly on the ground eating roots, young leaves, bugs, and berries. The forest canopy disappeared and opened the sky to view. The sun was a direct presence in Harun's life. Grass grew as the overhead branches and leaves thinned dramatically. The grass blades were surely inedible and their undomesticated grain seeds were likely indigestible as well. Much of the social learning from Sara's generation was useless by Harun's time and was already forgotten. Harun had to learn new things such as how to find food hidden underground.

Fortunately for him, the learning of Sara's day could be replaced much more quickly than her genes.

Change as radical as the move from forest to woodland is very dangerous. Surely it led to the extinction of many species. Harun was lucky that apes are intelligent observers, opportunistic in their feeding tastes, and flexible in their actions. If figs were no longer abundant, roots were there to be dug up. Yet Harun is an interloper in a strange setting. He can survive on the ground, but not very well. When he digs into the earth, his fingers are not well adapted to the task. When he tries to pull up a root, he has a hard time gaining leverage. Many of Sara's traits serve Harun like an ill-fitting set of clothes. Survival for Harun's generation means relaxing natural selection's conservative grip. Varieties of hands that would have failed in the past as not best suited for picking fruit might pass muster as being better at digging, or at least no worse than old-style hands.

No evolutionist looking at Harun peering into the earth could predict that somewhere at the end of this process of change would come speech, yet it is apparent that selection could no longer favor the old ways. Something new was bound to come.

> An ape from Harun's troop lies dead beside a termite mound. A very young ape sits on top of her; apparently it is the dead mother's child. Harun is moving away from the corpse and pays no mind to the orphaned young one.

Perhaps most shocking to twenty-first-century humans is the pitilessness of Harun's world. Living in a world without mercy, Harun shows none himself. When disease takes an adult female, as happens routinely in the course of a year, there is sometimes an infant survivor. Perhaps it is still nursing or is still too young to live independently of its mother. Harun knows some of these young, has seen them play, and has perhaps played with them a bit himself, but, of course, he is never tempted to take charge of a little one and its lost future. The doom of young ones was nothing to wonder at, or even to regret. Disease, predators, and internal competition all take their tolls. In a human community, somebody—a relative, friend, or neighbor—would likely take in an orphan, but that sympathy is not so common in apes. Yet it is not quite impossible. Christophe Boesch, who spent over thirty years observing wild chimpanzees, reports cases of "adoption" (providing food and care for at least two months) of unrelated, orphaned young. So perhaps Harun will turn around, go back, and pick up the orphaned infant.

Harun sits well above the ground in a thorn tree. He has built himself a small nest for sleeping. The sky to the west is a deep red.

Chimpanzees and gorillas spend much of their time on the ground, but they return to the relative safety of trees for sleeping. It is almost certain that for a very long time the woodland apes still returned to the trees at night, even if they spent most of their days on the ground.

Alisa | 4,500,000 Years Ago | Generation 125,000

A two-legged ape stands erect and tries to get at the roots of a medium-sized bush by pulling the whole thing up from the ground. This ape looks something like Sara of a million and a half years earlier, but she is taller and her body has been modified to walk upright on the ground. Her teeth have gotten smaller, especially her canines. And, just to show that this ape is not Sara, Alisa is presented with gray-and-white coloring. In truth, of course we don't know what any of the hair coloring of any of these fossils were like, just as we do not know the colors of fossil birds or dinosaurs. The woodland setting in this diorama is dryer than in the previous display. The grass is more sparse and more like standing straw.

We have jumped far ahead in time because our focus is on speech and even its rudest beginning came millions of years after Sara. Notice, however, that although Alisa is still far from being human, she is no longer like Sara. Many people who object to evolutionary accounts of natural history distinguish between what they call microevolution and macroevolution. The change between Sara and Alisa is an instance of macroevolution, the appearance of a new species with new adaptations.

Microevolution is the sort of biological change that happens so quickly nobody can deny it. Examples are all around us. Flu viruses mutate and become immune to previously effective antibodies. Insects became immune to pesticides. Changes like these have happened in only a few years, or even less in the case of the flu, so it is unusual for a creationist to insist that such changes are impossible. Instead, they often say, "Oh, sure that is true, but that's just microevolution. It's macroevolution that I don't accept."

The evolution of a new species happens over long periods of time and produces the sort of gross changes that turned arboreal Sara into the mostly ground-dwelling, bipedal Alisa. Many small changes—what a creationist would call microevolution—took place and were preserved, and then more changes built on those older changes. This long process of changing details building on details is what creationists call macroevolution. Biologists do not make that distinction because macroevolution is just evolution that has gone on for many generations, enough to transform a tree-dweller into a walker, or even a mute creature into a speaking one.

> Beside Alisa stands a half-grown juvenile who bends over to pluck a mushroom from the earth. In the background are two acacia trees. A pair of giraffes reach up to the branches of one of the trees. In the other tree stands another member of Alisa's troop, probably catching beetles that crawl along its limb. Still further back, against the horizon, is a volcanic mountain with just a bit of smoke rising from its flattened peak. Another member of Alisa's troop sits off a bit by herself, looking in the direction of the smoke rising from the volcano.

Fifteen thousand centuries after Sara and Harun, the pre-human lineage has become partially bipedal. The fact that walking on two feet came first is now well established by many fossils, but paleontologists previously expected that the first step to being human lay in acquiring a big brain. The theory held that we got bigger brains which made us smarter, which got us making tools, which made our hands busy with carrying and manufacturing so that we started walking on two feet. Fossil and archaeological evidence now proves the process went in exactly the opposite direction. Some people have tried to explain the reversal by showing that each step could have led to the next, but that approach is a guessing game and probably confuses causes. As a rule, important evolutionary changes are adaptations to the environment. There is no reason to suppose that walking on two legs necessarily led to tool use which necessarily led to bigger brains, etc.

The change to erect, two-legged walking is one of the great mysteries of human evolution. Many woodland animals occasionally stand on their hind legs. Vervet monkeys are four-legged animals that sometimes walk on two legs for short distances. Baboons walk on four legs, but they commonly sit erect, using their hands like any human seated at a desk. They also sometimes walk on three legs while carrying something in their free hand. A woodland gazelle in modern Kenya, the gerenuk, is routinely sighted standing on its back legs to reach high leaves. These examples show that it is easy to be upright without being bipedal. Carrying, standing upright to see, feeding on top branches, using one's hands—all of these tasks can be accomplished without redesigning the skeleton to support bipedal walking. So why is Alisa bipedal?

Tempting as it is to guess what lies behind our bipedal history, the answer is unlikely to bring us any closer to understanding why we talk. Indeed, the reasons seem unlikely even to tell us anything about what it means to be a human today. The rise of bipedalism remains an unexplained fact of the fossil record, and so as long as we cannot say why our ancestors began walking upright, we will be missing something fundamental about the path from ape to human.

Alisa's species was possibly a known fossil species, *Ardipithecus ramidus*, of which we have an almost complete skeleton, but we cannot be sure that this exquisite fossil is a direct part of the human lineage. Sara's descendents likely adapted in a variety of different ways, creating several species and ecological niches. Except for us, the chimpanzee, and the bonobo, all those other lines eventually came to dead ends. Many accounts of human origins seek to describe the whole bush of prehuman and related species. But because our focus is on speech origins, our tour is different. We ignore most of the bush to follow our own lineage in Africa as Sara's descendents move toward speech.

One obvious detail indicating that Alisa and her companions are not yet ready for speech lies in the female looking in the direction of the smoking volcano. Among talkers, we would expect the observer to say, "Hey, that volcano is giving off smoke." Of course, without language Alisa's companion can say nothing of the sort; however, she could make a noise that catches Alisa's attention and then point or nod with her head toward the volcano. She does not. In fact the kind of *psst* and head-pointing behavior common among talkers is unknown in the animal world. As the French linguist Jean-Louis Dessalles notes, animals can communicate all sorts of things about themselves or their environment, "but they have never been described as drawing attention to events whose sole property is that they are unusual or unexpected." While humans compete to be interesting, other animals never even give it a try.

> Alisa and her youngster are lying flat, perpendicular to a creek, while cupping their hands to gather drinking water. The place has green grass and looks much more lush than the previous scene. Further downstream, kudu antelopes with grand corkscrew horns drink. The creek runs into a deepwater lake. A few ducks float in the water, but it is too deep for the shorebirds that are familiar to modern tourists in eastern Africa. Beyond the lake's far shore rises a steep cliff.

This habitat is very different from the previous display, yet Alisa lived in them both. Starting 4.7 million years ago, and persisting off and on for the next 3.8 million years, eastern Africa experienced a series of extreme climates known as wet/dry phases. The "dry" refers to the extensive drought that covered most of the region while the "wet" defines the patches of lush, watered lands in parts of the Great Rift Valley. This rift area is a deep crack in the earth created by ground motions, and by Alisa's day the rift already extended from modern Ethiopia into Kenya and further south. Extensive volcanic activity persisted during this long epoch and poured dust and ash over the landscape, changing the terrain as different features rose and fell.

Adding to the complexity of a wet/dry phase was the way wet areas shifted location over geologic time. A typical wet/dry phase lasted about three hundred thousand years. The one in Alisa's time had begun two hundred thousand years earlier, 4.7 million years ago, and persisted for another hundred thousand years after Alisa, ending about 4.35 million years ago. During that period a wetland might endure for fifty thousand years and then dry up, while another wetland would appear someplace else far away. For the species living through such changes, this pattern meant that from time to time they faced sharp adaptive stresses. First, there was a strong contrast between the wet and dry regions. Some groups adapted to one region or the other, but some moved between the two. Those middle groups had it rougher than the ones who enjoyed the lush lands year-round, but they may have been forced to be more cooperative and sharing to survive. Some work with simulators suggests that when the living is lush there is no reason to become less selfish and when it is too hard there is not enough to make sharing pay, but in the middle ranges where the living is hard but not monstrously hard, cooperation is the best strategy.

These wet/dry phases were a kind of species pump, altering habitats and putting repeated adaptive stresses on all the life forms. To survive in such a time, individuals had to become quicker learners than their rivals.

> The wall displays a large time chart that marks the points where different species of Sara's descendants appeared and disappeared. At first the comings and goings appear to be random, but there are shaded areas stretching across the time line. They mark the wet/dry phases. Looking more closely at the chart, we see that the births and extinctions are all associated with the beginning and end of the wet/ dry phases.

If chance alone were at work, we would expect only a third of the species to be associated with a wet/dry phase. The oldest *Ardipithecus* fossil yet known, for instance, dates from 4.6 million years ago, a hundred thousand years into the first of the wet/dry phases. Approximately 4.5 million years ago there was a deep lake along the equator in Kenya's Rift Valley. A bit later a soda lake appeared in the Afar region of Ethiopia. *Ardipithecus* fossils have been found in both places. The last known fossil dates to about 4.3 million years ago, when the wet/dry phase had ended.

All the changes visible in Alisha's life and form took place an eon before there was a hint of speech. Nothing was foreordained by either bipedal walking or the onset of wet/dry phases, but there was a coincidence of new things under the sun. The march of change persisted.

Yikin | 3,200,000 Years Ago | Generation 233,333

Deep time is a mystery, made doubly breath-stopping by its simplicity. One million three hundred thousand years passed between Alisa's generation and Yikin's. It sounds straightforward, but it is beyond imagining. All those little bits of destruction and creation that accompany the sweep of the second hand in shallow time seem to change nothing fundamental, but in deep time those changes amount to something profound and lasting. They redraw the landscape and the life forms embedded in it. The terrain of Alisa's time was so utterly forgotten by Yikin's day that the two seem to live on different worlds, and yet there are threads of descent connecting the two ages. Like relay runners connected by the passed baton, Alisa's generation became another and another until Yikin's turn to grab and run came.

During the million and a third years that followed Alisa's generation, two more complete wet/dry phases came and went. The first began 4.2 million years ago and lasted about three hundred thousand years. It was that period that provided the oldest-known *Australopithecus* fossils, the first genus of prehuman apes to radiate into different settings across eastern and southern Africa. *Australopithecus* success produced a variety of bipedal species and demonstrated the new group's ability to take advantage of the changing habitats. The claim is disputed, but most paleontologists agree that the australopithecines were, for a time, our ancestors.

Another wet/dry phase began about 3.65 million years ago, and it too lasted about three hundred thousand years. During this later wet/dry phase, several new *Australopithecus* species appeared, and there was also a new genus called

Kenyanthropus. That wet/dry phase was also the time of the famous Laetoli footprints.

> A photograph shows an eighty-foot-long set of upright prehuman footprints found in solidified volcanic ash, looking for all the world like tracks left by people on a beach.

The footprints are unusual for the way they capture the motions of a moment. The volcanic dust that contributed so much to shaping the wet/dry phases had just fallen like fresh snow and the creatures of the area had to tramp across it.

Normally such footprints in the ashes of time would blow or wash away, but in this case more lava came down and preserved the footprints the way a hardened concrete sidewalk sometimes shows the paw prints of a passing dog. The lava prints show that the bipeds of that moment in deep time had lost the thumb-like grasping big toe of apes and the prehumans of Alisa's generation. They could no longer use their feet to grip branches in trees.

> On a moonlit night, lions feast on a buffalo they have killed beneath a tall thornbush. Other buffalo can be seen in the far background, but no members of the human lineage are visible.

If our ancestors could no longer nest easily in trees, where did they spend the night? Sleeping on the ground would have put them within easy reach of predators, and nights are long in equatorial Africa—about twelve hours long, with little variance during the year. There is no evidence that they developed any ability to act in the dark, yet they survived. We have no idea how. Did they stand guard at night? Use old warthog burrows?

Create shelter from thornbushes? Climb back, a bit clumsily, to nest in trees? One guess is as unsupported as another.

In dry brush country, two upright males stand near one another making cackling sounds. One of them delicately plucks a louse from the other's head while making a cackle. The two are in a dry brushland. Trees are further apart than in previous scenes and the ground has foot-high dry grass, but the terrain is still far from the open grasslands common to modern Africa. Bushes with long thorns stand in all directions. Yikin and his friend have almost no body hair and their skin is thick and gray, like rhino hide.

This simple setting portrays one of the main moments in the emergence of speech: the rise of voluntary, or semi-voluntary, vocalizing. Semi-voluntary means that there is an urge to do something, but it can be controlled if an individual wants to suppress it. Apes have the ability to control many actions, but apparently their vocalizations are rarely voluntary. In this display, Yikin and friend are grooming and making cackling sounds, but they are not just like cats purring. They are beginning to take control of a new kind of action.

Why the change? Many things have happened since Alisa's generation. The adaptation to ground dwelling continued. By Yikin's time the head was located much more directly over the spine. The typical four-legged mammal's neck tilts forward, making it easy for the head to look directly ahead. The neck bones enter the skull at its lower rear, angling the skull to look straight ahead from a tilted neck. For two-legged creatures, the tilted neck and angled head would make an individual look naturally toward the ground. Overcoming this problem, the human neck is upright and the spine enters the skull directly at its base, making the head as erect as the neck and trunk. The entry point between neck and skull is well demonstrated in many fossils, and by Yikin's time the head was set straight over the spine in full bipedal mode. The lineage had become committed to walking upright on the ground. Although the change opened the possibility of penetrating new habitats, it turned the line away permanently from the treetops its ancestors had occupied for tens of millions of years. The three-dimensional navigation that was required in the trees became sharply reduced, perhaps freeing up some brainpower for other things.

Many paleontologists have noted that *Australopithecus* legs were short and stumpy, like tree-dwelling orangutans, instead of the long striding legs of modern humans. Some interpret this feature as an indication that despite the lava-frozen footprints and pollen data about the habitat, these creatures must have spent considerable time in trees. But David Carrier, a biologist at the University of Utah, has proposed an alternate explanation: Short legs lower a person's center of gravity and make it harder to knock them flat. Keeping on one's feet is important for a species that regularly fights over access to females. *Australopithecus* males likely did compete for sexual dominance, as suggested by their skeletons, which were quite a bit larger in the males. Carrier's explanation seems plausible and fits the other data about *Australopithecus*.

Ground life seems also to have led to sharply thinned hair, although that proposition is more controversial and unattested by any fossils. Some scholars date hairlessness as recently as one million years ago, but the DNA evidence says otherwise. That evidence comes from louse DNA. Body lice are common parasites of many primates, and they plague both chimpanzees and gorillas. Each primate species has its own species of louse pest. Chimpanzee lice, for example, can be found anywhere in a chimpanzee's fur and are specialized for moving through chimpanzee hair and drinking chimpanzee blood. Humans are unusual in supporting two species of lice on a single body—one adapted to life on the human head, and the other adapted to living in human pubic hair. This presence of two species on a single host is possible because they live on separate hair "islands" and do not compete directly with one another. The study of louse DNA indicates that our lineage has supported two louse species for the past 3.3 million years. Hence, we know that by Yikin's time the body hair had reduced dramatically.

The pressure to reduce body hair is mysterious; its causes are as much a source of speculation as bipedalism. It was not, however, as rare as upright walking. Some large woodland animals such as the rhinoceros, elephant, or hippopotamus are relatively hairless. The display shows Yikin and his contemporaries with a thick hide like those other hairless beasts. Many anthropologists believe that modern human skin with pores for sweating was a much later adaptation to life on the grasslands and open savanna. So if Yikin's skin was not sweaty like ours and not furry like an ape's, what was it like? And why was it hairless?

Whatever the reason, the reduction of body hair was immediately revolutionary, more so even than walking on two feet because the costs were felt right away. One of hair's functions is crucial to holding primate societies together. Primates establish personal social bonds by holding long grooming sessions—picking insects out of one another's fur. Grooming allows primates to spend time together as equals and to form bonds that enable one another to pitch in together on other mutually useful tasks. Grooming is so critical to the social organization of primates that a retreat of lice to a few hairy islands would seem to threaten social disaster.

The anthropologist Robin Dunbar has proposed that language—particularly the ability to gossip—arose as a substitute for grooming. Other scholars sometimes scoffed at the idea as overkill. It is true that modern humans often entertain themselves and strengthen social bonds by gossiping, but good gossip requires a knowledge of grammar, a moral code that can be violated, and an interest in one another's outlandish behavior. None of these things were available when speech began. Yet Dunbar's primary point about the importance of grooming is an excellent one. Forgetting everything else in his theory, the fact remains that the loss of body hair forced our ancestors either to find some social substitute for grooming or to become less social.

Humans can form strong bonds without speaking. All they have to do is vocalize, making sounds without using words. Parents and babies, lovers in bed together, and fans of sporting events can all vocalize and form bonds without speaking words. Singing is another way to form group bonds. A group joining together in the meaningless *tra-la-la* of song can feel satisfied and unified despite the absence of either grammatical or moral rules. What's more, social vocalizing does not require the introduction of yet another unprecedented behavior in our evolutionary story. Elephants, whales, and dolphins all use specialized vocalizing to maintain contact over distances. Dolphins coordinate some group activities by making sounds, and elephants use very low-pitched sounds—far too low-pitched for the human ear to detect—to signal their presence beyond a herd's range of vision. Thus, the introduction of a new system of vocalizing as a grooming replacement was a familiar biological trade-off.

The origin of vocalized bonding is not as mysterious as two other human peculiarities, hairlessness and upright walking. Apes and monkeys are a noisy group, so some vocalizing during grooming is commonplace. As hair began to disappear from the bodies, this incidental vocalizing grew more important. The only part of this story that is particularly controversial is the human fondness for choral activity.

Harvard psychologist Steven Pinker argues that our taste for music is a "cheesecake" effect. We like cheesecake because our taste buds work that way, not because cheesecake is a nutritious food. Pinker says our taste for music is the same; we like it because we like it, not because it benefits us. His argument, however, is based on assertion and analogy, and, unlike cheesecake, a taste for music has not been shown to be unhealthy. Pinker's thesis inspired many music psychologists to find genetic and neurological evidence that our taste for music is no accident.

In the diorama, Yikin and his companion are not truly singing. They are in the earliest stages of semi-voluntary vocalizing. Their throats were ape throats, and we have good fossil evidence indicating that, like other apes, Yikin's generation still had laryngeal air sacs. The best they could do at this stage of evolution was likely make the *oo* and *uh* sounds of a three-month-old infant. However, the display does not even use these human sounds. We shall never know how Yikin's generation really sounded, but they were more likely to sound like a normal monkey howl than like a syllable or collection of syllables.

Ape calls are "holistic" and carry no special meaning within their parts the way a subtle change in syllables can change the meaning of a word—e.g., *confirmation* and *conflagration*. It takes an especially alert hearing system to catch these distinct, internal sounds, and this first step toward speech cannot have demanded more of the auditory system than it was prepared to handle. Evolution is often a two-sided process in which multiple details change in a synchronized way. For some reason, body hair begins to thin, but that process can only work if something makes up for the weakened ability to form social bonds. So in comes voluntary vocalizing. Yet vocalizing can only work if the ear starts paying closer attention to the sounds each vocalizer makes. This need to coordinate many different changes is one reason evolution is so slow. It took over two million years for the body to adjust itself fully to walking on its hind legs. Many distinct bones had to be altered; muscles had to be tailored to new stresses; society had to adapt to the new frontal display of genitalia; a life in the trees had to be foresworn. Those changes took many scores of thousands of generations. And Yikin shows that a similarly thorough revolution has begun with the introduction of voluntary social vocalization.

> Yikin and four others from his troop are sitting in the shade of a tree, turning over bits of gravel and snatching the insects they expose. In the background is a soda flat, a dry basin that was once a lake and now supports no growth larger than microbes. In the foreground, lying

beside one of the female prehumans, is an infant. The mother makes a
steady flow of coughs for the infant to hear, and the infant coughs back.

An alternative idea about how voluntary vocalization got
started comes from anthropologist Dean Falk. She suggests
that as prehumans grew less hairy, their young faced a serious
survival problem. Infants could no longer simply cling to their
mother's hair and get a free ride in the manner of monkeys
and apes. The solution, Falk proposes, was to put the baby on
the ground and keep it in auditory and emotional contact by
vocalizing. We can wonder whether the grooming-replacement
idea or the baby-on-ground suggestion is better, but there is no
reason to choose between them. Evolutionary changes are often
overdetermined, with one solution easing multiple pressures.
Ape fur supports many social needs, and the most efficient response would be to satisfy as many of those needs as possible
with one new, neat, and easy alternative. The change to infant
vocalization does not appear to be a difficult evolutionary problem. *Australopithecus* already made some vocalizations. Getting
an infant to vocalize seems to be chiefly a matter of lowering the
age at which individuals make these sounds; in other words, the
timing of an existing trait's onset was adjusted. Nothing had to
be started from scratch, nor did a series of separate traits have
to be recognized in the way upright walking demanded.

There was likely a long time when even an active australo-
pithecine toddler could not keep up with the troop. As their
mother's hair became too thin to hang onto, adults had to
carry the young, and, if the adult had other pressing things to
do with its hands, it had to put the baby down. An infant on
the ground is in a very dangerous position, even if it can walk.

Predators are an obvious threat, but so are forgetful adults. A distracted adult might wander too far from the toddler and lose sight of it. A frightened adult might flee a predator and, in the terror of the moment, forget the child. To defend against such forgetting, a toddler needs to keep reminding its mother of its presence. Making simple sounds is a way to make contact, although not a risk-free one because the sound can attract predators. The risk of abandonment is too great, however, and even on the African plains where hyena and other predators abound, a young grazer, like a wildebeest, begins making a regular call sound if it has become separated from its mother.

An argument against the baby-on-the-ground idea might be that modern babies are about three months old before they get much beyond crying, but we can never assume the past was exactly like the present. The long, helpless infancy of modern infants reflects the enlarged brain that still has much growing to do. Three million plus years ago the prehuman brain had not yet begun its dramatic tripling in size; the stresses promoting prolonged infancy were absent.

This difference in development rates and stages is why we cannot simply look to modern children and conclude that the first human-like vocalization was an *oo* since, after crying, it is the first vocalization made by modern infants. A common error in thinking about evolution supposes that the growth of a fetus and then of a child reflects the whole evolutionary history of the child's lineage, or, to put it in jargon terms, *ontogeny* (growth of the individual) *recapitulates phylogeny* (evolution of the line). That idea does work sometimes, but more often it is not quite exact and we can never be sure ahead of time whether this rule of thumb will apply or not. It takes some extra evidence, reducing the ontogeny evidence to secondary, supporting material.

The limitations on the throat and hearing, the use of holistic sounds, and the ape-sized brain all make it hard to believe that the original, voluntary vocalization was any more than a set of simple, distinct sounds. Thus, we have no evidence to suggest that the sounds of Yikin's generation were like the baby sounds we hear today.

> An acacia tree spreads over a brushland. One large leopard climbs the tree's trunk carrying its prey in its strong jaws. The prey is Yikin; his left arm is held tight while the body hangs freely. Only the carcass's dead weight resists Yikin's ride to the leopard cache where an adult gazelle already hangs.

Anne | 3,000,000 Years Ago | Generation 250,000

Twenty-five thousand years after Yikin's generation disappeared, still another wet/dry phase (the fourth of the series) began. With the return of this contradictory weather pattern, the human lineage once again had to keep adjusting and adapting, adapting and adjusting. One hundred and seventy-five thousand years after its start, this wet/dry phase was still underway and the human lineage kept up a persistent pattern of adjusting and then adjusting again.

The effects of these wet/dry phases are not well understood, but as a general rule there are two ways to adapt to environmental change: physically or behaviorally. Physical adaptations require new genes, so some genetic shift would have been required every time climate change brought environmental shifts that made physical adaptations obsolete. Behavioral adaptations call for a change in habits, and individuals can react much more quickly than genes if the habit is based on intelligence. Thus with the wet/dry phases, we should see many comings and goings of physically adapted species, and we do see that process in the fossil record. Behavioral adaptation should select for the ability to respond intelligently to new circumstances, and there are many intelligent animals on the African savanna—elephants, hyenas, wild dogs, and perhaps many of the birds. But we do not know how the intelligence of these species has changed over the past four million years as they too survived these recurring wet/dry phases.

> Anne stands in shallow water, and a dead flamingo hangs limply in her right hand. Her left hand is held up before her, in a defensive gesture, while a larger male from her troop snatches at the flamingo. There are no other flamingos to be seen in the lake.

This diorama depicts the kind of behavioral adaptation that might have taken place. The shoreline and shallow-lake birds—spoonbills, ibises, herons, storks, and others, as well as flamingos—are more tempting than catchable. In photographs or when viewed from the window of a Land Rover, the soda lakes of the Great Rift Valley can seem like glimpses of Eden, but they are more like a devil's playground. First, the water, for all the beauty of its reflected sky, is corrosively alkaline, so astringent that flamingos need special defenses to stand in it for hours. Most animals do not like to set foot in a soda lake.

Then there is the heat. This whole evolutionary story takes place either directly on the equator or within just a few degrees of it. The altitude kept the temperatures tolerable, for the most part, but the soda lakes can approach 130 degrees Fahrenheit these days and might have gotten still hotter three million years ago. The sandy soil surrounding the lake bed holds the heat right on its surface, burning any feet that walk on it.

Also, there is no vegetation to draw animals. The alkaline basin kills off the grass, so grazers have no reason to come. The flamingos have a special ability to stand in the lake and feed on the water's microbes, but with all the flat, open country surrounding the lake, no predator has any hope of surprising a water bird. Maybe a young male lion—alone and desperate—will make a stab at hunting them, but he soon learns the futility of the project.

Yet the birds are only birds, and from time to time one gets sick or injures itself. An alert killer, scanning a lake with a hundred thousand flamingos in it, might notice one that is in trouble. The hunter could then walk directly toward the lake, enduring the equatorial sun as it tramps onto the scorching shore and finding no relief as it splashes into the poisonous water. All the birds but one take flight and escape, but the troubled one that caught the hunter's eye cannot fly to safety.

> Anne and two other females sit very close together beside a tiny stream that flows slowly toward a distant, shallow lake filled with flamingos. The three are making different sounds. In front of Anne sits a toddler entertaining itself by pulling on grass blades and shouting, "kah . . . kah . . . kah." Another toddler of about the same size sits in front of one of the other females, poking its hand in a hole in the earth while chanting, "moo . . . moo . . . moo."

In our history the critical response to changing conditions is the adaptation to a loss of body hair by replacing grooming with vocalizing, and also vocalizing to keep contact between children and adults. This account of speech origins contradicts a popular alternate theory that supposes speech was preceded by signing, the use of hands to form words. That argument's chief supporting fact is that apes make voluntary, illustrative gestures already, seeming to give them a straight path to making words. Another argument comes from infants who can learn signing a month or so before they can speak.

Our scenario takes a different approach because signing by itself does not resolve the problem of hair loss and the attendant emotional bonds due to grooming and vocalizing on the ground. Nor does it explain why we ever did acquire voluntary control of vocalization, or how speech became the default medium of expression. (There are no hearing communities whose primary means of communication is through hand signs.) Illustrative gesture, of course, is also a normal part of speech. "Turn left," a helpful person says and illustrates by moving the left hand. And facial expressions are part of it too. "Ugh, what a smell," says a child while ostentatiously holding its nose. And surely, from the beginning, language has combined face, gesture, and voice.

The presence of gesture among apes is thought provoking, but does not imply that sign-only language was a mere step or two away. When apes gesture, they communicate to control. Human language has an informative use unknown to ape societies. The illustrative go-left gesture, for example, is functionally very different from the controlling gimme gesture. Going from gesture to language, thus, was not just a question of stretching an existing capability, but adding a new function to the repertoire. And there seems to be no way of predicting how an animal will evolve a completely new function.

Another bit of evidence supporting speech over hand signs comes from two Chilean neuroscientists, Francesco Aboitz and Ricardo Garcia. They report that the brain's "dorsal auditory pathway is specialized in vocalization, processing, [and] enhancing vocal repetitions," and sign language takes advantage of the same dorsal auditory pathway. The suggestion is that language production, vocal or gestural, emerges from the brain's sound/hearing region.

Perhaps rather than making gesture-based, communal communication more likely, the existence of gesture-based, controlling communication made the route less likely. After all, voluntary gesture already had a function, a function that persists to this day. Humans find it very easy to read chimpanzee gestures because we have them too. Yet when it comes to the primal cooperative gesture—pointing—chimpanzees do not understand. Nor do they understand a go-left gesture. As Michael Tomasello put it, "You cannot tell an animal anything."

> Anne and two other females are walking and vocalizing as they go. Anne and one other female each carry an infant pressed against their shoulders. The third female is childless, but she carries a stone.

To be clear, Anne is still a bipedal ape, but in this diorama she and her companions are doing something unknown among other apes. Functionally, they are acting more like birds—not songbirds, who learn complex chirps to attract mates and hold territory, but more like migratory birds, who make steady sounds, particularly at night, to maintain contact with one another as they fly. This kind of calling depends on lung power instead of the subtleties of vocal-tract control that songbirds

and speakers use. And it seems clear that our vocalizations began using the standard ape lung capacity. Even so, it is important to realize that something new has appeared in the primate
order. With the loss of hair, infants have become a persistent
burden. All primate youngsters depend on their mothers for
survival, but children that must be deliberately carried require
much more active support than mama chimpanzees need to
indulge. With this demand for active support, the human lineage is developing a need unknown to the other primates and
is developing a solution unknown to them as well. They had to
build on their capacity for making sounds.

A Japanese primatologist, Nobuo Matasaka, has studied the
difference between infant and monkey sounds and found that,
to this day, the sounds that infants make in the second month
are within the capacity of many primates, but human caregivers teach them how to turn those vocalizations into the sharing
of an emotion. Mother and infant then take on an accented or
rhythmic quality common to both music and language.

Another American linguist, Judy Shepard-Kegl, who was
present to observe the creation of Nicaraguan Sign Language,
reports that the ability of deaf schoolchildren to create a new
sign language rested on four inherent qualities, the first of
which was a love of rhythm or prosody. Rhythmic interaction
preceded words.

And anthropologist Dean Falk reports that mother and
infant interactions during the first months of infancy produce
a "melody arc," that is, a "single rise, then fall of pitch in a
sound . . . in one expiration of air."

These observations all point to one theme: Caregiver and
infant develop the foundation of language (and music too)

by forming a rhythmic, emotional interaction. The vocal use of patterned sound to express emotions is older than speech. Thus, it appears that between the gesturing ape and the talking *Homo*, our lineage went through a unique period of vocalization that was more familiar to birds than other primates, but the function was a replacement of the old primate actions of grooming and clinging.

> Anne is part of a group of six walking upright through brushland. Among the six is Anne's older brother, and Anne's toddler sits with legs around the brother's neck. Both of them are chanting, "*kah...kah...*"

Changing the means of carrying out a function like traveling with an adult while preserving the function itself is almost certain to produce some novel results. In this diorama the stable function is transporting a child that is too small to keep up. The new means is carrying the child rather than letting it simply hang on by the hairs. This solution added a new degree of freedom: Carrying the child is no longer forcibly limited to the mothers. Naturally, fathers in Anne's time had no idea who their children were or even that they had children, but maternal uncles knew their sister's children, and thanks to the cooing sounds, uncles also had the opportunity to develop some emotional rapport with the infants. The pressures of the persistent burden keeps the maternal bond strong and probably even raises it, but there is also a compensatory strengthening of the bonds with others so that the mother is not left, as with most primates, to bear the burden alone. Cooing infants can form accidental bonds with nearby adults, who, thanks to the sounds, can recognize individual infants at a much earlier age than was previously typical. Thus, while most primate mothers

fiercely keep their young to themselves, the balance changed in
the human lineage. Mothers became somewhat less jealous and
others felt more connected to infants. The change permitted
novel behavior like an uncle carrying his sister's child. When
people think of our own evolutionary history and what separates us from other animals, we usually concentrate on the big
brain and its reasoning powers. But it is scenes like this one,
of a child on an uncle's shoulders, that get to the essence of
the difference. By three million years ago, we were beginning
to show signs of emotional entanglements built by vocalizing.

Mahasti | 2,700,000 Years Ago | Generation 275,000

At dusk, a red sky hovers over thornbush country. Two cheetahs, one
of them only half grown, are looking up from a partially eaten impala
whose lyre horns are slightly bloodied. Striding toward the group are
six hairless, upright apes. As they come they make a rich variety of
whoops. Three of them hold stones, which they might throw when
they get closer. Slightly apart from the other upright apes comes
Mahasti. Instead of carrying a stone, she holds a dead infant whom
she is unwilling to abandon.

By Mahasti's time, the human lineage had been using vocalizations as a means of bonding for over half a million years. In
this scene, that effect has become two-sided: making the vocalizers feel the strength of their group while alarming outsiders
with the sight of a unified challenge.

Three details in this scene reach back into deep time: the
cheetahs at a kill, the mother carrying a dead baby, and the yelling. Many a primate mother has clung to her dead newborn,

giving it up only when it becomes too smelly to endure any longer. The behavior probably combines a bit of instinct, emotion, and some recognition of what has happened. Yelling in the presence of predators is also very old, observable to this day among chimpanzees in the wild. The novel part is the use of loud shouts as an aggressive action that does not merely defend against predators but turns the tables on them. The shouts remind us of the scene in Sara's day when apes faced off in the branches making cries, but those howls were emotional while these suggest a cooperative plan.

Africa's brush country is not silent. Elephants are noisy when alarmed and zebras make a barking sound when they get excited. Hyenas produce a weird, legendary cry that does not sound much like laughing, but is truly strange with its unsteady, high pitch. Animal chases, however, are mostly silent. Hyenas can chase zebras or wildebeest for miles without either the predators or prey making any kind of cry or howl. So the introduction of an action that uses vocalizations as a tool to alarm and confuse is something unusual, although lions too can use a good roar to panic a herd into the jaws of waiting lionesses.

The evolution of technology eventually saw three phases—tools, secondary tools (the use of tools to make other tools), and machines (tools with parts); not surprisingly, this scene is still in the first phase. Tool use was very old, and stone-throwing is just a form of tool usage. We might wonder why earlier primates did not also throw stones. One answer that has long appealed to popular imaginations is that these "tools" are really weapons and humans became humans when they began using weapons, but that explanation seems melodramatic.

Some apes include meat in their diet, and presumably would be willing to make life easier with a weapon or two, if they could. Indeed, apes in zoos are often reported to throw things at people, and a chimpanzee in a Swedish zoo was filmed stockpiling stones and throwing them at visitors. But apes do not throw objects very far or with much force. Anyone watching the action of a chimpanzee in a zoo throwing stones, dirt, grass, and even feces is likely to say that the behavior is more like a particularly elaborate dominance display than like the use of a weapon.

The weakness of the chimpanzee throw points toward the more plausible explanation for why earlier tool-using apes did not take to hurling stones. Even though modern apes are on the whole much stronger than modern humans, the wrist action required for effective throwing was a late evolutionary change that occurred only along the human lineage. Apes use their shoulders or elbows when trying to introduce force into their hands, so they have neither the wrists nor finger grip to pitch a rock in a way that might consistently do damage.

The dates for the rise of the human lineage's ability to throw stones is disputed, with some people claiming evidence going back beyond four million years. Thus, the scene used in this display of stone-throwing less than three million years ago is not particularly controversial. (But keep in mind that nothing about human origins is without controversy.)

The reason for the increased power of arms, wrists, and hands along the human lineage is also a source of disagreement. Some argue that our ancestors began to throw things with the clumsiness of an ordinary ape, but then selection favored the better throwers so that in time we got the hands and arms to do the job right. That reasoning is not much different from the classic argument of Jean-Baptiste Lamarck that the giraffe grew a long neck because generations reached toward the high branches. Lamarckian arguments run from function to form. We imagined a function (throwing well) so we evolved a form (the redesigned arm) that could handle it.

Darwinian reasoning commonly runs in the opposite direction, from form to function. We had arms and hands flexible enough to enable us to throw things, so we began hurling stuff. How did we happen to develop arms capable of throwing? Presumably it came through a selective process that supported existing functions like pulling up roots and tearing at scavenged flesh. Eventually the arms that were better adapted to these existing functions were powerful enough to support throwing as well. This emergence of unintended consequences is a regular theme in any account of Darwinian evolution, and distinguishes it from the older theory of Lamarck in which results were the logical consequence of purposeful behavior.

Mahasti and another female sit together at the edge of a thorny clearing watching a lioness and two cubs about a hundred yards away tearing at a gazelle. The other female has placed an infant on the grass. It lies calmly, making simple sounds. The two adults are making complex melody arcs.

The two females are not conversing, nor are they having mock conversations by taking turns. Instead, both are making long strings of sounds, like *mabolipah*. These long nonsense sounds are not so common today because most children by the age of thirteen months are using words, but late bloomers, who even at fourteen months are not using many words, prove their normality by making elaborate babble sounds, just as in the case of the adults in this display. And twin children raised together are sometimes seen making babble sounds back and forth in a joint emotion. Complex nonsense sounds are also found in song and in religious ecstasy. Long melodies are an excellent way to give rich voice to one's social emotions.

The idea that the human lineage spent as much as a million years improving its ability to vocalize voluntarily without getting around to talking might seem incredible to people living in an era when things change so rapidly that in two decades fax machines went from cutting-edge to old-style technology.

Human history has moved on wings, but our prehistory crept on caterpillar legs. The Neolithic (or New Stone Age) period, when many large mammals became extinct and agriculture was first tried, began only ten thousand years (five hundred generations) ago. Since then, we have gone from a species of tentative planters to one that has connected all the world's computers into a global network. To us, the idea that our ancestors were still not talking thirty-plus thousand generations after taking voluntary control of vocalization might seem fantastic. But remember, Mahasti and her generation were still not human, not even protohuman. Their brains had not gotten any larger. Their vocalizations had not yet produced any kind of oral culture. Most of all, their society appears to have still been a typical ape system in which a troop faced the world as a unit, but faced each other as competitors. They were changing, adding a more active dependence on the group, but nothing decisive had yet happened.

The pushes behind biological change come from the changes in climate and vegetation, but that kind of change comes only occasionally, even in an epoch of wet/dry phases that bring lush oases to an otherwise persistently dry and difficult terrain. The previous wet/dry phase had lasted another fifty thousand years after Anne's generation, and then the climate gave way to another long dry period. A new wet/dry phase was just getting underway during Mahasti's generation. Very slowly, the whole African terrain was getting drier and cooler, but that process had been underway for the entire period since Sara's time. Eventually, the change would fundamentally alter the African ecosystems, but those events still lay in the future. The unforested areas were still woodland, often quite open woodland, but not yet the wide grassy plains. Mahasti was marvelously well adapted to this thorny setting, but she still seemed no more than that: an animal well suited to her habitat.

A dead female lies on the ground with a youngster standing on top of her. Mahasti is bending over to pick the young ape up.

And here at last we see an important change that is not based on climate, but the workings of emotional bonds. We said earlier that adoption has been seen rarely among apes, so this display does not depict something unprecedented. Instead, the balance is shifting as emotional awareness expands. Thanks to vocalization, Mahasti had already responded emotionally toward this infant when her companion was alive, and the emotion did not die with the death of Mahasti's companion. This diorama depicts another bit of accidental bonding that could arise from the introduction of infant vocalizations. The ease with which vocalization allowed unintended bonding kept reaping new consequences.

One of the major biological differences between humans and apes is that humans are much more fertile. Even among the most traditional, least medicalized of human peoples, the infant survival rate is much greater than among apes in the wild. Also, human mothers do not wait as long as chimpanzees do to have another infant, a startling fact when you consider how much more rapidly a chimpanzee matures. The fundamental difference between ape and human mothers is that ape mothers do their work alone and unsupported. Human mothers have community support, and in this diorama where Mahasti picks up an otherwise doomed infant, we see the beginnings of that change to community support.

Ling | 2,400,000 Years Ago | Generation 300,000

Ling and two other males from his troop stand together, holding hands, in open, flat grassland. No landmarks are visible except that, far in the distance, stands a mound of yellow-orange barren rocks called a *koppie*. Much closer, almost on top of the three bipeds, is a solitary hyena. The three are staring directly at the predator, even leaning slightly in its direction, while making a kind of low rhythmic melody. The hyena, momentarily flummoxed by the group's determination, stares back.

In a save-yourself world of individuals, these three are standing together to save themselves as a group. Primates have long demonstrated some actions oriented toward others, like grooming and consoling somebody with a hug, but those activities carry very low risk. Standing to confront a killer raises group support by an order of magnitude.

Hunter-prey standoffs on the African grassland are not unknown, but they are rare. A mother zebra or wildebeest will stand against a solitary hyena that has discovered her newborn. Buffalo will sometimes gang up on a nearby lion pride. Elephants do not need to attack; they just form a line and the sheer bulk of their display sends all but the most foolish onlooker scampering. Smaller prey species, however, including apes and monkeys, respond to predators by running away or scampering up a tree. But in this scene there are no trees in sight, just open grassland. Africa's multimillion-year process of drying and cooling has finally produced a dramatic change in habitat. Far to the north, well beyond the knowledge of any creature in our history, the periodic rise of ice-age-strength glaciers has begun, while fossil pollen shows that grassy plains have started to dominate the equatorial East African terrain. For the woodland animals, this transition to Serengeti-style open savanna was as challenging as the loss of forest had been three and a half million years earlier. Members of the human lineage had to survive in the grassland without being able to outrun any of the plains predators. Another challenge on the open grassland is finding enough to eat. Grass is normally not edible. Most of the grazers on the grassy plain have complex stomachs that permit thorough digestion of grass by softening it before the animals regurgitate the cud and chew it some more. Primates do not have that kind of stomach and are in serious danger of starving on a savanna.

The prehumans that evolved in the woodland became bipeds, shed most of their hair, and then started to vocalize. It made them quite distinct from their founding ancestor, Sara, but they were adapted to bush country, not open grassland.

Many prehuman groups did not survive the transition beyond the woodland. It seems that *Australopithecus africanus*, after a seven-hundred-thousand-year run, went extinct under the new pressures. A later *Australopithecus* species, called *garhi*, also disappeared. The *Australopithecus* genus, after 1.7 million years, had almost run its course. In its place came two new genera: *Paranthropus* and *Homo*. From its start, the *Homo* line has been distinguished from the other bipedal apes by a larger brain, smaller teeth, and a more delicate jaw. The new mouth indicates that *Homo* did not adapt to the grasslands by eating the grass. Instead, the teeth suggest they turned to eating more meat; however, they did not develop the claws and fangs that are typical of meat eaters. They had found a new way to survive in their desperately difficult habitat.

In this tour we have spoken more of the human lineage than of a particular genus or species. In most cases, we do not know for sure which species was part of our lineage and which was a collateral branch on the prehuman bush. And even if a particular species was included in our lineage, it is likely that at some point the line split and our lineage went one way while the older species continued on for some time further before it died out. However, the record is clear that with the rise of the African savanna two and a half million years ago, there was an abrupt shift in the bipedal populations. From this point onward in our tour, we are observing members of the *Homo* genus.

Does that mean they were human? If you met one on the street, you would probably say no. They did not talk, they still had very small brains, and they had no technology. (Although scratches on fossil bones do indicate that by that time they were using stones to smash open bones and get at the marrow.)

Yet it is also true that they were no longer apes (unless you are one of those who like to say that humans too are apes). Look at what the trio in this scene are doing—standing together and facing a deadly threat as a unit. We often think of technology as the secret of human adaptation, and it is important, but even more fundamental is the willingness to stand together and cooperate.

If the threesome were ordinary apes, each would probably run, but these three are making a stand, keeping themselves together with their hands and their precursor of song. Could such a tactic possibly work?

Lone hyenas in a face-off with prey are cautious. They look for an opening and dart in, trying to make progress through quick nips. The prey's task is to keep facing the hyena, preventing those flanking nips. The hyena might suddenly swing around slightly to aim, say, for the standing biped on its left. The alert trio is excited and might speed the tempo of their noise as they pivot to keep the hyena centered, facing them directly. The hyena is much faster than the trio and can run in a full circle, while the bipeds can only spin on their heels to stay facing the predator, then spin on their heels again as the hunter continues in its circle. The trio's sound might become louder and more energetic. When the hyena stops, the threesome could lunge forward a step or even two, looking straight into death's face. They are so close that a punch in the hyena's snout seems possible.

If the three are lucky, this standoff could last for hours, until the hyena gives up and lopes away in search of less troublesome prey. Mothers of newborn wildebeest have sometimes been observed making such a long stand, saving themselves and their calves by outlasting the hyena. So the trio in this stand might survive, but why would all three stay together?

From a plain Darwinian perspective, the calculation is simple. If all three bipeds run, one of them is almost certain to die. If they stand together, one of them may still die, but there is also a chance that all three will survive. To use fanciful numbers just for purposes of illustrating the point: If one tactic (running) leads to a 1 percent chance of all three surviving, while the other tactic (standing) leads to a 10 percent chance of all three surviving, evolution should favor the second tactic.

But evolution is never quite that straightforward and predictable. Ling might believe that he can outrun his two companions, so if he runs he has an 85 percent chance of surviving the encounter. But if he stands, he has a 30 percent chance of dying. Perhaps, then, Ling should decide it is every man for himself and start to run.

Most of the animals that herd together on grassy plains follow exactly that logic and run. Gazelles, zebra, wildebeests, eland—they all run. These are the save-yourself species and natural selection keeps them speedy. The slower runners get eaten. For Ling and his companions to stand there, taking their chances together, selection has to favor a completely different solution. This time, selection favors the most cohesive group. A whole new set of traits is favored at this level. Group members standing as one and facing a predator together need courage, fidelity, and an affection for the others who stand beside them. None of those traits are necessary at the level of individual competition, and none of them are conspicuous in the ape world.

But talk of group selection is likely to surprise readers who have not yet heard that multilevel evolutionary theory has made a striking comeback. Their astonishment makes sense. Single-level selection—also known as selfish-gene theory, reciprocal altruism, inclusive-fitness theory, or kin-selection theory—was given a mathematically solid foundation in a series of papers published by W. D. Hamilton in 1963 and 1964. A dozen years later Richard Dawkins published his classic *The Selfish Gene* and put the gene-level idea of evolution on everybody's lips. Meanwhile, the idea of group selection had become overused and fell into disfavor. Many people forgot that Darwin had said that selection worked on both the group and individual level.

The biggest challenge to the theory of single-level selection probably lay in explaining ants and bees, most of whom toil their whole lives without reproducing. These insects serve in many a fable as the great example of individuals who sacrifice themselves for the good of the group. E. O. Wilson's classic study *The Insect Societies*, however, showed that through a peculiarity of their reproduction method, the sisters in an ant or bee colony are more closely related than they would be to their own offspring. By supporting one another in the colony, these sister insects do more for the survival of their own genes than they would if they had gone off and had children of their own. This discovery did not resolve all questions, but it so neatly reduced nature's most impressive example of cooperation and self-sacrifice to gene selection that the selfish gene became the starting point whenever a question of evolutionary explanations arose.

The insistence that all selection takes place at the gene level also produced a grave problem for human origins. Single-level selection seemed to push human history outside of evolution. It became common to hear that people had somehow freed themselves from evolutionary forces. The reasoning was simple: If selfish genes are the whole of evolution, we must be outside of evolution because selfish genes cannot explain familiar human behavior. Adopting unrelated children, giving directions to strangers on the street, passing the bread at a table, doing unto others as you would have them do unto you—none of these things can be explained in terms of selection at the gene level.

Not surprisingly, an effort to explain all human behavior in terms of selfish genes did take shape. Perhaps its most popular claim comes from putting a scientific veneer on an old pre-selfish-gene rhyme about the war between the sexes:

Higamous hogamus:
Women monogamous;
Hogamus higamous:
Men are polygamous.

Charming as the observation is, it explains neither faithful husbands nor faithless wives.

Human speech has proven especially difficult to explain with selfish genes. Why do we share information with non-kin? Gene-level altruism does such a poor job of explaining it that we might look in the opposite direction. Perhaps speech arose as a tool for deceiving our neighbors. It is a selfish enough guess, but offers no clue as to why anybody should listen to a race of liars. Others suggest that speech promotes status in the group, and it does, but only after it has evolved well beyond the level of simple statements. Gene-level competition is hard put to explain the rise of the speech triangle. Why do people reveal their secrets? Why do people listen? And why does anyone bother to pay serious attention to neutral topics?

The most complete effort to understand human speech in terms of selfish genes comes from the French linguist Jean-Louis Dessalles. His suggestion is that an individual with a reputation for honest, relevant speech will have a high status within its group, although he admits that a social tool like speech in a world run by selfish genes is a paradox, and he does not address all the other altruistic elements of human society. Efforts to insist that the primal cynicism of selfish-gene theory can explain human biology have probably done a great deal to revive the anti-evolution spirit that grew after the 1960s. It was as though science had endorsed the maxim of La Rochefoucauld that "we would often feel ashamed of our best deeds if the world could see all the motives that produced them." The remark passes muster as a literary *mot*, but as a scientific statement about the basis of human behavior it is mighty narrow. Selfish motives might explain the cutthroat politics of the academy, but it does not offer the slightest clue as to why some people risked their own lives to hide Jews from the Nazis.

Thus, for humanists, it was something of a relief in 2005 when E. O. Wilson surprised the world of insect specialists by reporting that examination of the genetic structure of the most cooperative insects, which are called eusocial, showed that the closer-to-sisters rule was not always true. Some eusocial insects are not closely related, while some closely related insects are not at all cooperative. The link between kinship and cooperation is not statistically significant. At a stroke, the strongest evidentiary pillar for the universality of the selfish-gene theory collapsed . . . which does not mean that the gene dimension itself collapsed. Plainly, gene selection explains much; it just cannot explain *everything*.

In 2007, E. O. Wilson and another noted evolutionary biologist, David Sloan Wilson, explored the ground where they presently stood in an essay titled "Rethinking the Theoretical Foundation of Sociobiology." It was a mild-sounding title until you remember that E. O. Wilson first coined the word "sociobiology" and gave the field its original theoretical foundation. Essentially, one of the world's most distinguished biologists was saying he had changed his mind and now believed that "multilevel selection theory (including group selection) provides an elegant theoretical foundation." Single-level accounts of evolution were no longer the only assumption guiding inquiries into the evolution of behavior. Multilevel selection provides a basis for favoring moral actions (behavior that puts the group ahead of the individual) while keeping humans within the domain of biology. Of course, the multilevel idea remains controversial and occasionally takes on a round of personal insults.

The trickiest part of the multilevel theory focuses on how a population moved from domination by individual selection to one where group selection predominates. Gene-level selection is a stable process and is not easily overturned. We humans can see that we benefit from our more-cooperative world. Presumably, chimpanzees would benefit too, but they are caught in what we can call a Darwinian pit, a trap in which the competition to survive gives the immediate advantage to cheaters. Any system based on cooperation distributes benefits to all and pushes members up the sides of the pit toward more cooperation. But then some individuals slack off, reaping the benefits of the group while not contributing their share of the burden. The cheating corrupts like a cancer, spreading its uncooperative genes and breaking up the cooperative system. Escaping this trap requires something more than the fact that cooperation would be a good thing to do.

We have seen that emotions help hold primate societies together and can push against the pit walls. Grooming is especially important for building these emotions and reducing tensions. The result is powerful enough to prevent individualistic tendencies from eventually tearing ape societies apart. For the eight hundred thousand years before Ling's generation, the human lineage had been vocalizing as a grooming replacement, and we have seen that this form builds even wider bonds. It enables more than two individuals to bond at the same time; bonding can continue while individuals do something besides vocalizing, and vocalizing forms collateral bonds with infants and juveniles who are normally too young to be included in grooming. A vocalizing group can also feel its own strength as it vocalizes while approaching danger.

Even before it advances to the level of speech with words or grammar, vocalizing provides a way to move out of the Darwinian pit and not slip back. As Ling and his two companions stand and face the hyena, they hear one another's presence through the sound of their melody and feel the strength of their group.

Ling stands in a shallow creek bed chanting a bit of melody over and over. A companion lies down with his face near the water and cups a drink in his hands. Ling looks out over the grassland, staying alert while the other one drinks his fill, unconcerned about any danger that may be approaching. Presumably, when he is done drinking the two will change roles and Ling will get his drink.

This display and the previous one depict a fundamental fact about *Homo* communities: Members must trust one another to survive, but trust by itself is awfully dangerous. Suppose, for example, that Ling suddenly notices a lion approaching from one hundred yards off. He should alert the drinker so the two can stand or flee together. A faithless guard, however, can slip away, leaving the inattentive drinker to distract the lion while the deserter escapes. If there is a genetic contribution to infidelity, that gene survives while the lion makes a meal of the trusting drinker and its genes. The immediate benefits for the genes of cheats, defectors, and faithless ones outweigh the group's longer-term benefits from mutual support.

Vocalizing solves the problem by getting around dependence on unverifiable trust. Ling is cooing as he stands watch. If he proves unfaithful and runs away, his companion will be alerted by the disappearance of the vocalizations. He can look up, see Ling running off, and go after him. If they both survive the coming lion, the companion will know Ling is untrustworthy and shun him. If others learn the same thing from their experiences, Ling's unreliability might cost him membership in the group. Thus, the tables have been turned. It is now better for individuals to work honestly with the group because they can be made to pay the costs of cheating without undermining the community as a whole. Now selection favors fidelity over cheating, and a whole new evolutionary pressure comes into operation.

Active dependence has produced something new—verifiable trust. Infants have no choice but to rely on their caregivers, but adults are different. Other primates do not have to trust their neighbors, and do not trust them. Human adults are unique in their need to trust other adults. Unique needs produce unique solutions, so it should not be surprising to see the human lineage introduce a new behavior—trust-assuring vocalizations.

Here we can see two levels of selection working together. At the gene level, any genes promoting fidelity and courage will be favored. Selection at this level is made by the group rather than by nature, because it depends on the group penalizing the unfaithful and cowardly. Natural selection now works at the group level. If Ling's group is less adapted—i.e., less able to cooperate—than its neighbor, it can lose territory and population and is more likely to die out altogether. Multilevel solutions produce very different societies from those supported by genes alone, but they are just as ferociously competitive with their neighboring groups, maybe even more so.

We have good circumstantial evidence that the human lineage moved out of the gene-level Darwinian pit a bit less than two and a half million years ago. That date will surprise many people who have assumed that group loyalty and morality came much later in the story, perhaps as recently as a hundred thousand years ago. The major argument for a more recent date is that it took a long time for our brains to reach modern size, and that archaeological findings indicate that body adornments like bracelets and necklaces came into use at about that same time. Our ancestors were finally smart enough to see the value of cooperation—or so goes the argument.

Pushing the date of community cooperation back to Ling's generation sends that whole assumption topsy-turvy: It was cooperation that gave us big brains.

First, it does not take much intelligence to cooperate. We already saw the example of bees and ants. As insects go, they are clever, but they still have only sand-grain-sized brains, and when the honeybee genome was deciphered it turned out that their hyper-social behavior did not require many genes. Biologists have also found sacrifices for the group at the level of microbes, and microbes seem intelligent only when compared to a stone. A kind of bacteria called the *Pseudomonas fluorescens* is even less intelligent than an ant and yet it cooperates. It lives in water and is known informally as the "wrinkly spreader" because of a kind of cellulose mat that it creates on a water surface. The mat protects against situations where oxygen is unevenly dissolved in the water. Since it costs energy to create the mat, it would be to the benefit of a selfish bacterium to cheat. That way a bacterium would get the benefit of the mat without sharing in the risk of its production. But the wrinkly spreader is a stable species, producing its mats for generation after generation. Somehow it got out of the Darwinian pit.

Spinner dolphins have been seen hunting together, chasing prey fish into a crowded bunch and then taking turns at feeding on the prey while the others continue to keep the fish bunched together. Waiting a turn to eat suggests real group solidarity. These dolphins also appear to have gotten out of the Darwinian pit. Thus, while many animals do live in a Hobbesian world of all-against-all, quite a number of species have escaped that fate and live in cooperative groups; the idea that humans are cooperative and trusting is not as radical as it might at first seem.

Further support for the early date comes from fossils showing changes in brain size. Ling's generation was right at the point where prehuman brains grew significantly larger than chimpanzee brains. Brains are "expensive," meaning they need ample energy. It takes almost ten times as much energy to support a unit of brain tissue as it does any other part of the body. So any increase in brain size has to reflect either a reliable increase in food or a trade-off that reduces some other bodily demand. For example, the gut can shrink, thus allowing more energy to go to the brain. Something had to happen to permit the *Homo* brain to grow.

The chimpanzee brain, however, was already big. Primate brains in general are bigger than the brains of most mammals of comparable size. By the time of the chimpanzees, the primate body had already been trimmed down considerably to support the ape's relatively large brain. Further physical trade-off seems unlikely.

Another possible trade-off is behavioral. Bigger-brained species might cut back on their activity, and chimpanzees do spend much more of their time moving about than do human hunter-gatherers. But that difference comes from the cooperative nature of hunter-gatherers. They support one another, while chimpanzees have to keep on the move looking for food. Thus, while this kind of trade-off is possible, it requires something further.

Still another kind of trade-off is in reproduction. Two Swiss-based anthropologists, Karin Isler and Carel Van Schaik, have found that relatively large-brained species have longer spaces between births, but chimpanzees already have a large space between births. Extending it even further could make the cost in reduced fertility just too great. Moreover, we know that instead of increasing the space still further, the human lineage developed shorter spans between births. Isler and Van Schaik report that "each lineage faces a 'grey ceiling,' i.e. a maximum viable brain size . . ." Beyond that point, the lineage's reduced fertility becomes so great that there is a serious risk of group extinction. There is, however, a way to poke through that ceiling. Cooperative child rearing can reverse the findings about mother-alone rearing. Instead of reduced fertility, the cooperative lineage can enjoy both bigger brains and increased fertility. Thus the sudden growth in brain size in the human lineage is strongly suggestive that this was the point when cooperative behavior began.

And where there is effective cooperation there must be trust. With trust, the motivation supporting the speech triangle becomes possible. In their essay on group selection, the two Wilsons argue that speech is a "fundamentally communal" behavior and it "probably required a shift in the balance between levels of selection *before* [it] could evolve. Only when we could trust our social partners to work toward shared goals could we rely upon them to share meaningful information." (Italics added.)

> Ling stands off afar, showing his back. In the foreground are two females. One, very pregnant, is squatting on the ground. The other female is holding the pregnant one's hand, ready to help with a delivery the moment she is needed. The female in labor is moaning; the attendant is making a soft sound. Ling is walking away, unaware that his child is about to be born, unaware indeed that he has anything to do with the mystery between women.

Our understanding of the rise of cooperation is still incomplete. Particularly mysterious is the transformation in the role of males. Human men have a reputation as testosterone-maddened, violence-prone, lady-loving cads, but compared with chimpanzees, bonobos, and gorillas, the human male is astonishingly domesticated. Apes have many different ways of organizing male-female relationships, but in all of them the males compete to produce offspring. Among gorillas, the competition is direct. Males face off to control female harems. Among chimpanzees and bonobos, the competition is more at the sperm level. Males and females engage in widespread sex without forming pairs, but the sperm engages in a hidden battle as it tries to impregnate the female and block out rival sperm. Meanwhile, humans also have a great range of organized male-female relationships, but except for the very rare systems in which women have multiple husbands, there is very little competition between males to produce children. It is not entirely absent. Evolutionary psychologists have sought out evidence for it and found some, but it is striking how scattered the examples are while in the ape world evidence of male competition leaps at observers from all directions.

Genetic evidence for this difference between ape and human males has also been identified in the Y chromosome. Only males have a Y chromosome, and it turns out that the chimpanzee Y chromosome is radically different from the human version. The chimpanzee Y chromosome has lost half of its genes for coding protein. Ape and human Y chromosomes differ in 30 percent of their DNA, suggesting that the Y chromosome has been changing much more rapidly than in the rest of the genome. For the most part, chimpanzee and human chromosomes vary

by only two percentage points. It appears that for the past six
million years, the chimpanzee Y chromosome has been evolving for still more effective competition against rival sperm. The
lead investigator of this research, David Page of Cambridge,
England, told the press that the Y chromosome has undergone
"wholesale renovation and remodeling." Exactly what this difference means is going to take much deeper investigation, but
clearly the human Y chromosome has not been undergoing the
same stripping for sperm competition. The human male has
diverged greatly from the ape male.

The finding came as a great surprise to geneticists, but for
anthropologists it was more like a confirmation of what was
already evident. Many anthropologists have noted that before
primate males could share or trust, they were going to have to
change dramatically from the normal ape. One anthropologist,
Terrence Deacon, proposed that our capacity to use symbols
has arisen in response to the need to organize ways to allow
males to act cooperatively in the community without getting
into constant fights over the females.

Meanwhile, a British team, Chris Knight and Camilla
Power, argued that females had to rebel against the males and
force a change before we could form the bonds of trust supporting language. These ideas are intriguing and carry promise, but they have very little fossil or archaeological evidence
that can either support or date the changes. For now it is one
more mystery. We have the Y chromosome data that supports
our observation that human males are very different in their
behavior from chimpanzee males. They are fathers, husbands,
uncles, and trusted advisers, but, like so much of human origins, how they came to be that way is clouded.

More generally, group cooperation can only endure for generations if the benefits of cooperation are shared equally. In Darwinian terms, results are measured in the number of descendants one has. That means not only equal access to sex, but also to food and protection so that everyone has an equal shot at living long enough to match others in numbers of children. This development is as fundamental as speech and must have evolved along with it, side by side, stride by stride. But we do not know much about that process yet and it is mostly invisible in this history.

Ng'ula | 2,100,000 Years Ago | Generation 325,000

Even though the weather had dried up about a hundred thousand years before Ling's birth and remained dry three hundred thousand years later, changes to the *Homo* line came quickly after Ling's generation. For the animals, this environmental stability meant that if a species was well adapted in Ling's time it was probably still well adapted to the environment during Ng'ula's generation too. For the *Homo* genus, however, there was a whole series of strong, new pressures as its dominant form of competition shifted from the individual-level to group-level selection.

We commonly think of group adaptations as behavioral, and there must have been many emotional changes to support the new behavior, but there were also anatomical changes. A good example of a group-supported difference between humans and all other primates is in the whites of the eyes. The sclera, as anatomists call the part of the eyeball surface beyond the cornea and lens region, is dark in apes and other primates, but is white in humans. In this case, we can be confident that the white sclera was not an adaptation to life on the savanna. Plains animals like wildebeest, zebra, gazelle, and lion all have dark eyes.

Like combinatorial vocalization, hairlessness, and upright walking, white eyes are universal among humans yet unknown in the rest of the primate world. White eyes let others in a group see where your attention is focused, forcing individuals to share what catches their notice. It is different for animals with dark eyes. As one anthropologist put it, "Like mobsters wearing sunglasses, mature apes are experts at being poker faced, displaying no eagerness at all for having their minds read easily."

After Ling's generation, *Homo* continued developing the emotions that support group loyalty. Change often comes in packages. A famous example comes from a Russian furrier who wanted to breed a fox that did not run away at his approach. After some generations of selective breeding he had his tame fox, only it looked like a dog, complete with droopy ears. The ears seem utterly irrelevant to tameness, but as the pharmaceutical industry has discovered, targeted chemicals are a bit of a myth. They always have side effects.

Our current knowledge about white sclera is too limited to say whether human eyes have been positively selected or if they simply came along for the ride. If they were positively selected, the sclera appeared as a mutant that helped the group and spread. If they were part of a general package, selection's role was reversed. If the white sclera were a handicap, as it would be for anyone competing chiefly on the individual level, we would expect a bit of fine-tuning in favor of some further mutation to keep the darker sclera. Thus, however it happened, the appearance of a white-eyed primate indicates something new: a creature that can prosper while sharing its focus of attention.

We have not advanced far enough in our history to set aside the third mystery, the so-called Harry mystery mentioned at the beginning of our inquiry: Why don't any other animals speak at all? The classic answer has been that other animals are not intelligent enough to use language, but that answer will not serve. Chimpanzees who speak not a word are more intelligent than intellectually disabled children who do talk.

Speech is an intelligent response to an unusual set of needs—trust and cooperation promoted by a relatively rare evolutionary pressure, group selection. The need for trust and cooperation is much more common among birds, especially those that cooperatively raise their nestlings, but birds, while much more intelligent than we usually credit them, are not bright enough to string words together.

How about parrots? Parrots, and crows too, might be smart enough to use some words, but they do not appear to be trusting and cooperative among themselves.

It is the combination of traits, each unusual in itself, that make speaking a one-lineage behavior. Primate intelligence is unusual, but not completely unknown in other lineages. Cooperation is found among birds, but is rare elsewhere, although not unknown. Trust is difficult to study empirically, but appears to be rare, although certainly within the capacity of dogs. Group selection too is rare, although it turns up across the biological universe. The odds against all of these peculiarities turning up together in a single lineage are very high, but by Ng'ula's time, 2.1 million years ago, conditions were ripe on the African savanna.

White-eyed Ng'ula sits on a pile of eroded rocks jutting up above a grassy plain. Beside her sits a three-quarters-grown juvenile who watches Ng'ula as she holds up a stone and makes rhythmic cooing sounds. Instead of smashing the rock against a bone, she is about to crack it against another stone. Next to her right leg lies a stone flake that has been given a sharp cutting edge. (Depictions of early tool use tend to show the first tool users as males, but there is no evidence to support this assumption.)

The oldest-known stone tools are flake tools, made about two and a half million years ago. They were fashioned by hitting one stone (the core) with another (the hammer stone). For best results, the hammer stone should hit the core at an angle so that sharp, cutting flakes break off at about sixty degrees from the angle of the blow. It takes experience to know how to use the hammer stone correctly, and the skill was almost certainly passed along the generations through some form of teaching. The watching juvenile will eventually learn by doing, but he gets the basic idea of what to do by watching. We have seen social learning of this type going all the way back to Sara pulling a vine while a youngster watches the method of getting fruit. But there is a difference. Pulling the vine is immediately rewarded, but what is the reward for chipping off a flake of stone?

The intellectual breakthrough here seems so simple that it is hard to credit it as a breakthrough. The artisan is using one tool (the hammer stone) to make another (the cutting flake). For millions of years before this moment, apes and other animals used tools to obtain a natural reward. Now, however, we see an intermediary step.

to use a tool
Use a tool ∧ to obtain a natural reward.

In mathematics, the process of cycling through steps while using the result of the previous cycle is called *recursion*. Counting, for example, follows the simple rule of taking the last number reached (N) and adding 1. Thus, if N equals 2, the next number is 3, and then adding 1 to that result you get 4. Another concrete example of a recursive process is (cycle 1) using a stone to make a stone tool and then (cycle 2) using that new tool.

Linguists are particularly interested in evidence of recursive thinking because their formal rules for generating a sentence often require a recursive process, although there is some disagreement over what recursion means to linguists. Sometimes it is taken to be a process of embedding one clause in another. For example the sentence "The apple pie came from Frieda's store" can be changed to "The apple pie that tasted so great came from Frieda's store." The embedded clause, *that tasted so great*, builds on a recursive process of reusing the first sentence. But, although embedded sentences are one example, recursive processes can also add clauses to the end—e.g., *Jackie threw the ball* ‡ *Jackie threw the ball to Pat*; or inserted at the front—*Jackie threw the ball* ‡ *Tom said Jackie threw the ball*. This ability to create longer sentences simply by taking the original sentence and adding a phrase at some point is an example of a recursive process.

The founder of generative linguistics, Noam Chomsky, has even speculated that the breakthrough to language came with the rise of recursive thinking. So when did recursive thought begin? It must have begun by Ng'ula's time because the use of a stone tool to make another stone tool shows practical, indirect thinking about how to use a manufactured object to make something still more useful.

> Four white-eyed females stand upright in a grassland. Scattered across the plain are small groups of zebras and a sprinkling of wildebeest. Toward the horizon a lone wildebeest staggers in a tight circle. Round and round it goes, round and round. The odd behavior is characteristic of animals whose brains are being destroyed by insect larvae. All four females are looking at the staggering wildebeest because Ng'ula points toward it with an outstretched arm and open palm.

An individual directing the attention of others by pointing is as close as we can get to the speech triangle without uttering words. A group has had its joint attention directed toward an object. Even though Ng'ula's ancestors had been vocalizing for a million years, speech may still not have appeared, but this display shows the speech triangle's immediate precursor. Somebody points and others look at a particular object. George Butterworth called pointing "the royal road to language." The joint attention and common interest of the triangle are both present. Still missing are the peculiarities of language that go beyond the speech triangle's bare outline, such as giving a name to a topic. The functional difference between a pointing gesture and a simple vocalization is minor. If Ng'ula had said, "Look," the result would be the same. The difference is in the promise for the future. As far as redirecting another's attention is concerned, pointing gets the job done, but what physical pointing cannot do is draw attention to an absent topic. If you want to draw attention to the topic of wildebeest when none are around, or you wish to compare something to a wildebeest, you need a name—a sound, a hand sign, a bit of scribbling. The pointing gesture will not do. Pointing is direct—*Look right there!*

Topics are permanently indirect. Someone can always respond to a proposed topic by saying, "So what?" It is a blunt way of looking beyond the topic and demanding a practical goal. It is no wonder then that other animals have no need for a topic. You can train an ape to do hand signs, but you cannot make him smart enough to see the value of placing an intermediary between this moment and his wants. Apes do not, so to speak, see around the corner.

The reason for the *Homo* breakthrough to indirect thinking is open to speculation. Was it because brains were getting bigger? Did it have something to do with group interaction? We can each guess, but the finding of tools made by other tools is clear. Prehumans had added a new quality to their thought.

> Ng'ula sits in a dried riverbed, nursing an infant. Her older son plays by cracking stones together. On the right, the riverbed's wall rises about ten feet. Ng'ula pays no attention to the thin stripes of ash, dust, and stone exposed in the wall. Her long-forgotten ancestor, Sara, was alive when a stripe toward the top was laid down. Ng'ula also ignores a group of yellow birds in a lone tree. They are hanging upside down and chirping gaily.

Shoichi | 1,800,000 Years Ago | Generation 350,000

We have been moving toward greater certainty about the species of our ancestors. At first, we could identify neither the species nor the genus with any sort of confidence. With *Australopithecus* we could be reasonably certain of the genus, although any particular species was more problematic, but by 1.8 million years ago we can be confident of a species in the human lineage. It was *Homo erectus*, the last of our ancestral species associated with the onset of a wet/dry phase. After a six-hundred-thousand-year absence, a particularly watery wet/dry phase appeared and large lakes stood in the Olduvai area of modern Tanzania, the Turkana region of northern Kenya, and at several places in the Ethiopian rift. As usual with a new species descended from Sara, *erectus* appeared at the start of this latest wet/dry phase. It might seem that with the abundance of lakes and the variety of ecosystems available, *erectus* as a group had many opportunities to find the best niches suited for them and to take on specialized traits adapted to those niches; however, the trend after two million years of wet/dry interruptions was toward flexibility, not specialization. Apparently *Homo habilis* had already spread as far as Indonesia, and *Homo erectus* would promptly spread out over much of Asia. *Homo erectus* had a much larger brain than *Homo habilis* and was surely the smartest creature yet to appear on earth.

> Three young bipeds are playing with stones. One of them, Shoichi, looks toward an approaching male, smiles, raises his hand and says, "*topoo.*" The two other children join in and call, "*topoo.*"

The first part of this scene does not mark any advance beyond Ng'ula's pointing at a wildebeest in trouble, and very likely similar associations between vocalizations and points of interest had been going on for some time. Associating a sound with a person is almost inevitable if you have a species with an urge to vocalize in the presence of others. The critical event in this diorama is the way the two other children join in.

One of the truly peculiar features of language is the way it depends on listeners to ratify a vocabulary. This detail is often missed by speakers who suppose that, like Humpty Dumpty, they are free to use words as they want, or that there is no such thing as incorrect speech. Self-appointed grammarians seldom make rules that stick, but the speaking community as a whole is perpetually negotiating details of usage such as nicknames and the borderline between colors. In this diorama we see an example of "word contagion." One child has an association between a sound and a person. The others pick it up and use the sound as a name.

Ratification ensures that language will only develop in usable, learnable ways. If Shoichi had unusual control over his vocal apparatus and said something unpronounceable by the others—something like *xkrbggc*—they would not repeat the word and make it their own. From the beginning, then, speech has been usable and learnable by the whole community without requiring special training. That simplicity makes speaking different from toolmaking, which has probably always required, as a minimum, specific teaching demonstrations.

At the start, speaking probably required only the ability to vocalize and form associations. In our own postmodern age, there are many words that do not spring from arbitrary association. A number like 1,462 is perceptually indistinguishable from 1,463 or even 1,429. We have such a number because we have a rule that generates it. Then there all those *tele-* coinages for sending information: *telegraph*, *telegram*, *telephone*, *telekinesis*, *television*, *telex*, and *telecast*. But these kinds of rule-made words cannot have been there from the start. The first words were random associations ratified by listeners.

Just how to translate these first words is also subject to dispute, and the challenge persists to this day. How best should we translate the word when an infant says *dada*? Is it *Uh-oh, it's daddy*; *Hi, daddy*; or *I see daddy*? And then there are the toy words that infants invent on their own and use for a time and maybe even hear echoed by the parents without there ever being a clear parent-language translation. Words in a full language are more sharply focused than the single-word utterances of protolanguage, and translations tend to be expansive.

The interpretation of single-word utterances almost always depends on both tone of voice and word, and often gestures as well. The tone contributes the quality of an interjection and can distinguish between *wow*, *oops*, and *goody*. The word itself names a thing or action. And a gesture can add illustrative information. For example, when a toddler cries *up*, its tone often expresses excitement or happiness, the word names an action, and a raised-arms gesture indicates a desire to be lifted up. Surely the first words of the human lineage—after millions of years of gesturing, and more than a million years of vocalizing—included gesture and tone, as well as a name for something, just as a modern child's first words do.

The introduction of names establishes the basic structure of language, the speech triangle. A speaker (Shoichi) and listeners (Shoichi's playmates) pay joint attention to a topic (topoo). From that perspective, this scene depicts speech. On the other hand, the diorama does not include the primal function of language, which in this tour is defined as considering a topic. Thus, this scene depicts a halfway moment when the social structure supporting speech exists, but not its function.

Some linguists have theorized that language must have begun fully formed syntactically, or at least fully functional, but modern infants and toddlers are also limited at first to this naming speech, identifying topics of interest without going on to consider them. By one year of age, infants are commonly speaking single words like *mama* and *dada*, and by two they typically put words together like *want juice* and *uh-oh fall down*. Typically, they only go beyond naming topics when they reach their third year.

A group of four adult males moves toward a distant suffering wildebeest going around and around. Shoichi, now fully grown, looks off to his left and sees, one hundred yards away, a second group of four bipeds. All of them are also looking toward that same circling wildebeest. Shoichi points with his hand and says, *"wahwahri,"* meaning perhaps *uh-oh, the wahwahri people.*

This display illustrates a persistent pattern. The young introduce a novel way of speaking, in this case using names, and when they grow into adults they use that novelty in a richer and fuller way; in this case the adult uses a name to spark a consideration of the topic. The difference is not in the speech itself, since in both cases, Shoichi simply combines tone of voice with a name and gesture, but in the response of the listeners. The children merely look and repeat the word, while in Shoichi's second diorama, the foursome are likely to respond by deciding as a group what to do. They will not be able to conduct much of a council as they do not have the language for it; at best they know a few words which they probably cannot yet link grammatically. Perhaps one of them will point to a ridge, or ditch, or some other inviting locale where they can wait to ambush their rivals.

Absent too is the brain anatomy that supports prolonged consideration of a topic; in particular, early *erectus* probably had yet to evolve the circuitry linking reflexive and deliberate attention into a functional unit. Without that system, words are likely simply to direct reflexive attention, leaving it up to each individual to use deliberate attention in silence. A council might go something like this:

1. A single word (e.g., *wahwahri*) catches reflexive attention.
2. Everybody uses deliberate attention to search for a solution.
3. Another word (e.g., *goody-ridge*) catches reflexive attention.
4. Listeners use deliberate attention to accept or reject a solution.

It is not yet what we can call language or a discussion, but in form and function our ancestors have started to talk.

Does this simple speech make *Homo erectus* human? That is a matter of debate and definition, but the presence of words and tools indicate that the human lineage had become richly cultural. If we take ordinary culture in its clumsiest sense of simply meaning the passing along of a learned behavior from one generation to the next, then chimpanzees and many other mammals have culture. Different groups of animals of the same species can have different ways of behaving in the same circumstances. But a rich culture in which indirect solutions—i.e., tools and symbols—are passed along from generation to generation is an exclusive part of the human lineage, and by the time of Shoichi's generation, such a culture was established.

We can see the difference between *erectus* and other animals by looking at the way it evolved its first words. In 2007, linguists and computer modelers met in Noordwijk, Netherlands, and, among other things, produced a list of three steps by which animals evolve a signal such as an alarm call or a honeybee dance. The first part of our tour through this gallery matched what the linguists found in the natural history of other signals, but then something happened.

The first step in the process that was defined in Noordwijk reads, "All signaling systems must begin as cases of accidental influence." In this instance, we saw that the accidental influence came from vocalizations that evolved as a way to overcome the loss of social grooming and the inability to hang on to a mother's hair. We saw that this behavior can lead to accidental bonding between infants and adults who hear their vocalizing.

The next step turns this accidental influence to the advantage of one party or another. As we saw, in this case the vocalizing infants gain a benefit by attracting more caregivers who can do things for the child. Examples of such benefits are picking up and carrying a child, or even adopting a child that has been orphaned.

The final step would be to expand the benefits so that, while retaining the multiple caregivers for the infant, the caregivers get something out of the vocalization as well. In our history, however, there was a change in function so that something completely new appeared. The contact between mother and infant on the ground and the bonding that comes from mutual vocalization are typical forms of animal manipulation, and even the formation of multiple caregivers are within the spirit of such manipulation. But speech is quite another thing, changing the function from manipulation of others to consideration of a topic. It is as unexpected as evolving a wing for flight and then using it to mark time.

Technically, such a change in function is called an exaptation, and, although exaptations are common enough in natural history, they do make hash of a simple one-two-three process of straightforward evolution. In this case, the transformation of human vocalizing into speech was brought about by having (1) sufficient intelligence to pursue indirect goals, (2) group trust that gave individuals the emotional entanglement and confidence to share what they know, and (3) a culture rich enough for members of a community to pass along indirect tools like hammer stones and words.

There were biological changes as well, most notably in developing a bigger brain and adding circuitry that supports controlled vocalization, but the key contributors were emotional entanglement, trust, indirect thinking, and purposeful culture. It was quite a remarkable change from Sara's time, a change that saw the rise of an unprecedented function: the willingness to share perceptions. And it was achieved by taking life one generation at a time.

CENTRAL HALL: BECOMING VERBAL

WONDER NEVER CEASES. We wondered about speech and found that it started when our ancestors transformed bonding vocalizations into a means of directing attention to a concrete topic. As a result, they were able to cooperate and prosper on the African plains. That explanation, however, only gives us something further to wonder about.

Every traveler knows there is a difference between being able to say a few words in the local language and speaking that language. Single words suffer from more than just a lack of grammar. They also lock the parties into a tunnel vision focused on a tiny bit of the here and now.

Pointing and saying the name of a rival group with an *uh-oh* tone of voice is a good example of the narrowness of a single word's focus. Contrast it with a full sentence: "Them folks across the plain are stalking the same wildebeest we are." After the blunt *wahwahri*, the full sentence might seem like just a long-winded version of the same idea, but the sentence shows how speech can work like a photographer with a zoom lens. Pull in tight and you have a group name, or zoom back for a panorama (*them, wildebeest, we*). The phrase "across the plain" even puts the scene into a physical context. And then we can combine sentences to describe many details of the scene—turning it into an episode. Today's speakers and writers of every natural language have their choice of where to set the zoom and what context to identify, but in Shoichi's time it was one word or silence. How did we get from words to sentences to episodes? That is what the second part of our history wonders.

It is tempting to say that once language got off the ground with single words, the rest evolved inevitably, just as the lone words of a thirteen-month-old baby foretell the coming sentences of the four-year-old child. However, grammatical thinking is different from the kind of thinking that can support one word, and a baby's brain has to do some growing before sentences are possible. The human lineage had to change as well.

Throughout this account, we have stressed that words pilot attention and that apes are perfectly capable of redirecting their attention in response to another. They can be trained to use single words in sign language or even to combine a couple of words into a valid phrase. At this level of language, their problem is that apes are not community-minded enough. Even when two domesticated chimps, both trained in sign language, sit together, they do not sign between themselves.

Could they do better if they became cooperative? Some scholars do argue that the difference between an ape's and a human's verbal ability is only one of degree, not kind. They point to trained apes and two-year-old children who share linguistic abilities—using one- and two-word phrases—and say that since there is no change in kind between human infant and adult, there must not be any difference in kind between ape and human adult either. It is a clever argument, but it overlooks the communal and syntactical blossoming that takes place in the human three-year-old. Something profound does happen in children at the start of the third year.

> Two small *Homo* adults stand and face a giant lizard known as a komodo dragon. The lizard is about seven feet long and holds its head at least a foot above the ground. The *Homo* adults are only about three feet tall with very low foreheads.

It is remarkable to think that perhaps as recently as seventeen thousand years ago there may have been a living species of talkers who used one- or two- or even three-word combinations but could not string them into a complex sentence. This was *Homo floresiensis* on the Indonesian island of Flores. It appears to have been a descendant of some very early *Homo* species, perhaps an offshoot of *Homo habilis*. It had a stone tool kit that included stones used to shape other stones. Somehow its ancestors left Africa and its line eventually found itself on an island only about five hundred miles from Australia. *Homo floresiensis*'s long survival without any growth in brain size— perhaps even with some shrinkage—shows us that the history to be told in this second part of our tour was not inevitable. Just because our ancestors had words there was no guarantee that paragraphs would appear some time later.

> A series of photographs are mounted on a wall beneath the words "Complex-Language Users." The first picture shows the novelist William Faulkner speaking at the University of Virginia; the second depicts a street rapper chanting into a microphone while strangers look on; and the third shows the actor Ralph Fiennes playing Hamlet.

While the *floresiensis* line was apparently static, our own lineage did change. Language became coherent and turned strings of words into an idea. Although apes sometimes put four or more words together, they have never managed to unite their strings into a full sentence. If apes today cannot organize their words coherently, we must assume the first speakers could not do so either. There has been some kind of qualitative leap between what today's humans can say and what early *Homo* managed.

Many people are, on general principles, skeptical of claims about evolutionary leaps. Evolution is not a ballerina. Species only need to be a little more fit than their competitors and do not leap far ahead in a single bound, but general principles can be fooled by details. In the case of *Homo*, competition between groups of speakers kept pushing rivals further beyond their old ape limits. Probably at no point did they make a leap—the kind of thing that makes grandchildren separate biological entities from their grandparents—but over thousands of generations the difference between them and their ape neighbors created the impression of a leap. Anatomically, we are unmistakably bipedal apes, but morally, aesthetically, and creatively, apes are more like mice than they are like us.

In the history of speech origins, this apparent leap takes the form of grammatical thinking and the capacity to speak in sentences. Sentences are coherent and frequently descriptive, and, in the classic schoolbook definition, they express complete ideas. This is not a bad definition, except for the part that leaves us wondering what makes an idea complete.

Where did grammatical coherence come from? We have no reason to suspect that maybe our ancestral line was just smarter about grammar than the other apes. Why would they be? Apes do display other kinds of practical thinking abilities like logical reasoning, associative learning, and perceptual knowing, but none of those kinds of thought depend on grammatical constructions.

> A photo shows the physicist Richard Feynman working at a blackboard covered in equations. Below the picture, beneath glass, is an old sheet of computer paper with perforated lines running up and down along both its sides and holes in its edges like on a reel of movie film. The paper itself has green and white stripes, and printed on it is a long section of computer code, written in Fortran, the earliest of the general-purpose programming languages.

Many people suppose that grammar is logical, but it is easy to form a sentence that is meaningful, grammatical, and illogical. For instance, the dramatic *He was a lion in her defense* defies logic. How could he have been a lion if he was a man? Of course, we could say *He was ferocious in her defense*, which seems more logical, but, hold on, what is that little word *in* doing? The preposition *in* refers to a spatial location or, metaphorically, a time period, but *her defense* has no space or time meaning. Most people, when asked to explain *in*, are baffled. Yet *in* is necessary to the sentence. Try dropping it. Suddenly the sentence loses coherence. Of course the sentence can be revised again: *He defended her ferociously*, which is logical, if boring.

The moral to this story is that while logical language is possible, the rules of grammar do not require it, and the rules of rhetoric positively flout it. Logical thinking and grammatical thinking are not identical, which is why mathematicians have invented their own systems for combining symbols with unambiguous reasoning. Experiments have shown that apes can learn to manipulate logical symbols. They are better at that task than they are at organizing words and phrases into ideas. Math is also distinguished by the way children require extensive, focused training before they become skilled at it.

Logical languages have been invented to run the computers of the world, and the fact that people can use these languages to control their machines proves that logical languages can be learned. The problem is that a line of code in a computer language is never descriptive, nor is it coherent in the way that sentences are. Instead of expressing complete ideas, these languages only issue instructions.

A cartoon on the wall shows two men, one very young, the other well into middle age. They are wearing ties and jackets that bounce as they smack each other's heads with rulers.

One discredited theory about grammatical expression is that it reflects purely associative learning. Extensive research has established that animals can be taught quite intricate behavior through a process of rewarding specific responses to particular stimuli. In the late 1950s, the leading experimental psychologist of the time, B. F. Skinner, attempted to prove that language could be learned entirely from its environment. Skinner's account stressed learned association between word and thing.

Skinner's theory is probably sufficient to account for the start of speech as depicted in this tour. Our ancestors vocalized for reasons that had nothing to do with language, but a few of those sounds became associated with people or things. When somebody needed to point out something, the association was there to use. Skinner's theory, however, was insufficient to explain grammatical organization.

Shortly after Skinner's book appeared, Noam Chomsky attacked its theory from many angles. He pointed out, for example, that language is so mysterious that it is not even clear what needs to be associated. Words seem obvious, but perhaps it is something bigger than the word, the phrase. Or maybe something smaller, like the -s that pluralizes nouns. If you favor words as the basic atom of a sentence, then a sentence is formed by combining words: The + red + ball + flew + by. Many linguists focus on phrases as basic: [The red ball] + [flew by]. Suppose I want the ball. We can add that information in many ways: (1) I wanted + [the red ball] + that + [flew by]; (2) [The red ball] + I wanted + [flew by]; (3) [The red ball] + that + [flew by] + was the one I wanted. Note that however you express it, the additional material goes between the original separate phrases. We cannot stick them between just any old word. How do we learn these elements? They seem to precede their associations; Skinner had no answer to that one.

Many people, of course, still believe that languages are learned and that associations play an important part in the learning. How else is a person to learn that the word *ball* indicates a spherical plaything except by association of word with thing? But something more elaborate must be going on when speakers use these associations grammatically.

An old fashioned device called a zoetrope is displayed on a table. It is a wide, slotted cylinder open at the top. Pasted inside the cylinder is a series of pictures of a ballerina. Slots have been cut into the cylinder beside each picture. If you spin the cylinder and look down, the pictures are simply a blur. But if a person bends down, bringing the cylinder to eye level, and looks through a slot, the ballerina appears to

be dancing. She is not dancing, of course; the illusion of the dance is
created in the head of the person looking through the slots.

We have said repeatedly that speech requires joint attention, and attention is a function of perception. So it might seem
that to know something perceptually is to know it grammatically, but such is not the case. Apes and many other animals
are masters of perceptual knowing. An ape swinging through
branches must see exactly where the branches are and judge
which ones are strong enough to support its weight. Its eyes
detect the branches, but its brain makes sense of the detail fly-
ing by.

An acrobatic poet might be able to describe the experience
of swinging from branch to branch, feeling the spring of each
limb as it pulls up and sags down, but describing is not doing.
Even a poet-Tarzan might be at a loss to say how he knows
which branches are safe to grasp. He just *sees* that they are safe.

Apes are experts at this kind of perceptual doing, while
completely lacking any descriptive capacity. It was, therefore,
no small feat when our ancestors moved from uttering single
words to speaking grammatically coherent, descriptive sentences. Many linguists consider the rise of syntax to be more
remarkable than the birth of words, but the primacy hardly
seems worth quarreling over. Words, phrases, and sentences
all make for wonder.

Knowing something grammatically is different from any
kind of knowing available to apes, and the story of this second part of our history tells of the evolution of that new way
of thinking. Remarkably, however, humans have several other

means of telling one another something besides using language and grammar.

A wall poster displays Kali, the Hindu goddess of time, change, and death, standing triumphantly over a dead woman. She points at her victim with one of her four hands, while behind her blazes a vast fire.

This single image captures all three ways that philosophers generally say we can tell somebody something: point, illustrate, or use symbols. The first and most elementary is pointing. Kali's outstretched palm leads the understanding eye. *There it is*, implies the gesture, making the simplest association possible with a topic. We already saw that our *Homo* ancestors were in all likelihood pointing for hundreds of thousands of years before they had names for their topics. Apes in the wild do not point, but captive apes will sometimes do it. If they became cooperative, as our ancestors did, they would likely be able immediately to start telling one another things by pointing.

A second means of telling is illustration, particularly the illustrative gesture. Instead of pointing, the illustration—also called a sign or an icon—conveys some intrinsic quality of its topic. The Kali image, for example, is full of details associated with death—a woman sprawled on the ground, a cobra, a fire. These images do not point; they show. You needn't know anything of Hindu mythology to see that something grisly is displayed.

Illustration is so powerful that even today our hands accompany and amplify speech. *Teeny tiny*, someone might say and press down with a hand to indicate just how absurdly small somebody was. Gestures alone are also sufficient in stereotyped situations: Moving the hand as though writing in air can signal a waiter to bring the check; pointing to a glass while catching a bartender's eye may signal the desire for another round. Many people have argued that such gestures or pantomime preceded language, although it is very hard to say something surprising about a topic through gesture alone. Gesture and words have probably gone together from the beginning of speech.

Apes are good associative learners and they already do gesture extensively among themselves, to beg for things or catch one another's attention. Once again, if they became cooperative, they could probably start using gestures to cooperate and tell one another things.

Yet there is more to the poster besides pointing and illustration. Many of the things in the image of Kali require knowing the culture. Why is Kali black? Why is she sticking out her tongue? What are the four arms about? Who is the dead woman, anyway?

For people in the know, these details are symbols that remind them of parts of a myth. For most of us, unschooled in Hinduism, these things are hopelessly obscure. But for those who love the story, the tale is visible before them. Religious symbols such as those in this poster stand for something else. Their meaning is not as self-evident as it is in pointing and illustration. Kali symbolizes death, time, change. You know that because you were told. Otherwise even that detail might be impossible to guess. Viewers need to share the artist's culture to understand what they are being told. Language is generally considered the most fully developed of any culture's symbol systems.

But we must be careful. Not every symbol works the same way. The Kali image is very symbolic, but its mystery is still grounded in something visual. Most words, however, take a further step by breaking altogether with any sensory grounding. In their most basic form, words are shared, ungrounded pointers. "Ungrounded" means there is no firm link between pointer and topic. A word is not like a finger, head nod, or eye gaze that locates an object in space. Nor is a word like an illustrative gesture that suggests the topic indicated. There is no natural link connecting the sound *shoe* with anything in the world.

Surprisingly, perhaps, this lack of grounding frees the word from the constraints that limits physical pointers and illustrations. Words work by common agreement. *Shoe* points to an imaginary shoe. If you don't know that, you're not stupid, you just don't speak English. And thanks to this common agreement, words do not have to point to something physically present. There are two reasons a named thing might not be physically present. Perhaps the word refers to an abstraction like justice, or perhaps it refers to a physical thing that does exist, like a shoe, but happens not to be present. Concrete words direct the attentive imagination as easily as the finger steers the eye.

This definition of a word as an ungrounded pointer arises naturally from our tour of speech origins. Language emerged from the practical need to direct joint attention toward an external topic. But that definition is very different from the usual philosophical definition of a word as a symbol standing in for something else.

This classical definition makes sense if you focus exclusively on a word's ungrounded, culturally agreed elements, but when you focus on functional roles—and function is what an evolutionary history must consider—there is no reason to assume that words function the way other symbols do.

A stop sign signals how to behave, and long ago Skinner proved that even pigeons can learn to act in accordance with such symbols. The same holds true for trademarks that proclaim an identity based on a set of associations.

A marriage rite changes community roles and duties, and there is nothing like a marriage in the animal world. Yet there are animal activities—call them rituals if you like—that do change relationships. Mating rituals are common enough, some are even charming, like the one in which a penguin brings a stone to its proposed mate. If she accepts, they are a pair for the season.

Linguistic symbols, words and phrases, do something quite different; they direct attention. Naturally, then, the workings of language become distressingly mysterious when we try to make generalizations that are equally valid for both words and symbols such as railroad crossings, baptisms, or arches of triumph.

All kinds of paradoxes emerge when we try to understand language this way. For instance, how does meaning fit into this symbolic accounting? Where exactly is meaning? And how does meaning differ from the symbol that supposedly represents it? The generalized definition of symbols has muddied the waters for thinking about language. If we keep the speech triangle in mind, and remember that words direct a speaker and listener's joint attention, the mystery of word meaning evaporates. The meaning of a word is where it points your attention.

Where, then, should we look for a clue as to what is special about grammatical thinking?

A glass case displays a series of skulls cut open to show models of their brains. On the left is a chimpanzee, with a brain at the back of its skull. *Homo erectus* has a brain twice the size of the chimpanzee's, and it is set on top of the skull. To the right is a modern human skull holding a brain three times the size of the chimp's and riding even higher on the forehead.

The most distinct fossil change during the past two and a half million years of our lineage was the increasing brain size, but most of our understanding of the brain and its complex actions, like supporting speech, is, to put it generously, hypothetical. We have, however, learned a few things, more than seemed likely only a generation ago. The ability to take pictures of what happens in the brain while it works has given us a mountain of image-based information. Doubters like to call this work neo-phrenology, after the pseudoscience that explained personality by studying bumps on the head. Yet scientists are beginning to map functional responses in the brain. It is slow, steady work. We still know next to nothing about the localized processes going on in those regions, and we have no idea how those processes become the sensations that define our experiences. Even so, mapping studies have been able to rule out certain ideas.

For example, imaging work has shown that using language depends on brain structures developed for functions other than animal communications. Groups of linguists continue to have bitter logical arguments over whether generating sentences depends on language-specific abilities or more general capabilities. Yet as far as brain-imaging studies are concerned, the answer is in. Language uses the same sensory areas used for other operations when we are not producing language. Syntax and semantics are not generated in special brain areas that go unused during periods without language. There may be specialized circuits that connect the separate words, but the foundations of these circuits are the sensory and motor areas that are used for more routine tasks. And many of these circuits do not appear to be inborn. People speaking different languages show vastly different patterns of brain activity.

Imaging work has also indicated that brain growth really did make our lineage smarter. An archaeologist, Dietrich Stout, directed a series of experiments in which his team taught people how to make different Stone Age tools. Then the team used imaging techniques to determine what part of the brain was used in making different tools. The images proved that the older tools depended on older brain areas. Oldowan tools, for example, predate the enlargement of the brain's frontal area, and sure enough, even in modern humans the frontal region is not used when making these basic tools.

Some of this brain growth might be a simple scaling-up of the existing ape brain. If a normal chimpanzee has a brain that can handle a few hundred words of sign language, an enlarged *Homo erectus* brain might be able to handle a few thousand words. A scale-up might also increase connections, enabling more efficient thought. Perhaps just by growing and adding connections, we became a more imaginative species.

For eons the brain has evolved through the progressive growth of brain matter, particularly the type called cortex, and by increasing connections. Typical is the change of eating and breathing actions. In vertebrates, both processes carry material from inside the face to the torso's interior—air to the lungs, food to the stomach. Both activities are critical and neither can interfere with the other.

The system of separating and maintaining the two functions was well developed by the time of reptiles. Both activities are controlled by a portion of the brain stem known as the reticular premotor region, one of the brain's oldest areas. All of a lizard's eating and breathing is handled reflexively.

The same basic facial structure and brain control is found in most mammals, with one change. There is now a connection running directly from the cortex to the old reticular premotor region, perhaps giving the mammal some sensitivity to its setting. Cows, for example, can look around before they start chewing.

Primates took even more control of their eating activity, adding direct cortical links to the tongue and facial muscles. As a result, monkeys can make many more facial expressions than a cow. Because of the cortical links, they can take more of their position into account while they act.

Humans have still more direct cortical control over their eating and breathing. In fact, we have a direct cortical link bypassing the reticular premotor area for every one of the controls. We have added direct links to the systems that regulate breathing and the operation of the larynx. These links do not mean we have gotten rid of the automatic operation of the reticular premotor neurons, but they can be overridden. Although direct control of the larynx and breathing has obvious benefits in shaping vocalizations, it also has a serious cost. We no longer keep our food out of our windpipes as automatically as the other species do, and food that is small enough to pass through the windpipe can end up in the lungs. This cost in choking is serious enough that the benefit had to be strong before evolution would tolerate it. Also, we disrupt our normal breathing patterns when we talk. We exhale slowly and inhale quickly, establishing a pace that would quickly make us dizzy in other circumstances. We have had to develop systematic ways of enduring these irregularities.

A particularly important case of enhanced regional coordination and efficiency in humans is the phonological loop. The story of its discovery parallels one of those accounts of a subatomic particle whose existence was first proposed by theory and then proven in the lab. In 1974, two psychologists, Alan Baddeley and Graham Hitch, proposed that humans have a "working memory" that demands deliberate attention on the part of its user. Part of the system works by remembering a sound, which enables us to remember what a person said long enough for a listener to retain a whole sentence. Working memory is also important in organizing one's thoughts into groups of words.

More than thirty years later, two Chilean neurologists, Francisco Aboitiz and Ricardo Garcia, announced that they had identified the proposed circuit. It links a region of auditory cortex known as Wernicke's area with a region of motor cortex called Broca's area. The auditory cortex enables people to think about sounds, as required by working memory. The motor cortex lets a person make particular sounds. By linking the two regions, the circuit provides a shortcut that lets a person turn the sound they are thinking about into moving muscle. Apes presumably think very little about sound, but if they ever do give it a thought they lack the wherewithal to turn their thoughts into vocalizations.

Aboitiz and Garcia call their circuit the dorsal auditory pathway. They say it is "specialized in vocalization, processing, enhanced vocal repetition, and short-term memory capacities that are crucial for linguistic development." Over a long period of evolution, the dorsal auditory pathway connected more and more areas, giving human sound-making "an unprecedented richness."

The ape brain was already a complex apparatus, something not to be tampered with lightly. When we see such complex changes to the *Homo* brain, scientists and nonscientists alike are a bit awestruck and wonder how it was possible to rebuild the ape brain. Most random changes would act like a stroke and simply undermine ape efficiency.

Terrence Deacon, a polymath at the University of California in Berkeley, and author of the invaluable *The Symbolic Species*, has proposed a way for brain change to happen without damaging what already is in place. Instead of a mutation or a shift in the balance between genes, a string of genes along a chromosome can accidentally duplicate itself. Duplication has led to some notable changes in body architecture. A centipede, for example, replicates the same body structure many times because it replicates its body pattern gene many times. In the primate lineage, a duplication led to the growth of extra material in the retina that was eventually adapted to enable color vision.

Deacon, who teaches a popular course in evolution, proposes that a section of chromosome for brain growth duplicated itself, providing extra brain matter. At first the change was neither helpful nor ruinous. The original brain matter was built correctly by the original genes and still worked fine. The extra growth was like a third hand poking out of the elbow, useless and horrifying, but not lethal. Over generations, we would expect natural selection to get rid of the duplicate, but there is a second possibility. Genes in the duplicate string can find something new to do; they can mutate and be selected for a modified function. Thus, the centipede's legs have come to work well, although they probably began a bit clumsily. A long-ago primate's eye began seeing colors. And, perhaps because of duplication, humans can do things well that an ape cannot do even poorly. It would seem then that when inquiring about human uniqueness, we should be alert for repeated, or segmented, thinking.

> A display case holds six axe heads fashioned from stone. They are identified by plaques showing dates—1.6 million years ago, 1.3 million, 0.9 million, two at 0.7, and one more at 0.4 million years—and where they were found, one each from Africa, West Asia, Turkey, and Central Asia, and two from Europe. They all look very much alike, and each has a cutting edge that has been worked symmetrically on both sides of the stone.

By 1.6 million years ago, *Homo erectus* had made a great stride forward in the creation of a much-improved stone cutting tool, the Acheulean axe. These axes were manufactured continuously until about a hundred thousand years ago. Technically, they are known as biface tools, meaning they are two-sided, unlike the Oldowan flakes. Anyone who has hefted a modern axe has seen the basic idea of the biface tool; the axe head has been worked on both sides to form a sharp cutting edge. We know with certainty, therefore, that early in the time of *Homo erectus*, the human lineage was able to pay attention to two sides of a stone at the same time and make use of what they imagined.

This tool indicates a great intellectual advance, and encourages one to suspect that there may have been some equal improvement in language. There are some arguments that the same part of the brain needed for precise control of the hands in shaping an axe is used to control speech, but it is hard to know what to make of such a fact. More important is that biface tools and true sentences both combine separate points of attention into a whole. With an axe, both sides meet at the cutting edge; in a sentence, two phrases share a verb. With the axe, the toolmaker looks at one face at a time, keeping the unseen side in mind so that the whole is understood. Similarly, a sentence like *Alphonse took the chair* focuses separately on Alphonse and chair; the two are brought together by *took*, a word that performs the unifying function of the blade's edge. The comparison works abstractly, but whether it has practical effects is another matter. Even if the Acheulean axe does suggest that *Homo erectus* had the intellectual hardware to link points of attention, there is always a lag while culture catches up to the use of its abilities. In our history there was quite a pause between the indirect creation of Oldowan technology and the use of another indirect tool, the word. The culture of early *Homo* was hyperconservative.

These theoretical matters are worth emphasizing because students of speech origins lose their way whenever they forget them. In the long march from Sara in the forest to Shoichi on the grassland, we kept our eye open for the rise of the speech triangle. With the introduction of words, this history reported a true biological novelty: a cooperative species able to share perceptions of a topic by directing attention with words.

It is almost as though our lineage developed a new sense, a cooperative sense that for the first time let individuals know what another sees, hears, or even feels. Texts about the biology of communication routinely open with a chapter on animal communications, but a chapter on animal perception would probably be more apt. As the linguist James Hurford noted, human language marks the latest chapter in the long effort of living things to know what lies beyond their skins. Remember that word *know*, which is linked to perception. The word was not *control*, which is the issue in machine and animal communications.

The emergence of a cooperative sense depends on effective social coordination and interaction. Joint attention leads to joint action. We can look at Shoichi's band using its first words and then take advantage of our long hindsight to see that much more power and understanding is yet to come.

How will it arrive? Through the speech triangle, by maintaining joint cooperation and attention to a growing number of topics, explored in greater and greater depth.

Christopher Columbus and a New World islander stand together touching each other's shoulder.

This scene commemorates a momentous event in world history—the reunion of far distant peoples and their languages. Humans went their separate ways perhaps sixty-five thousand years ago. By the standards of this history, that split was only yesterday, but it was still over 3,250 generations ago, enough time to invent writing, occupy the Pacific Islands, and plant the world's fields with grain. Some *Homo sapiens* stayed in Africa, others moved along the coast of Asia, while a third group moved north into central Asia. And from there the migrants covered the world. As the great churning and separations of peoples persisted, their languages followed remarkably different paths. That reunion in 1492 of people from the opposite ends of these long wanderings offers some insights into one of the most contentious debates about what happened during the long march from words to sentences. Columbus and the native islanders spoke quite different languages, and yet it was possible to learn each other's tongue. Much of what they said struck each other as manifestly absurd, as how could it not? Culturally, they were separate species—with different dress, religion, manners, arts, metaphors, tastes, and customs. Yet the relations of the speech triangle had endured and were immediately taken for granted by all parties.

European and American alike combined words into coherent strings and could immediately learn a few of each other's words by pointing and listening. Little is known of the grammars that Columbus encountered, but they were undoubtedly strange. The men and women of many Caribbean cultures spoke so differently that the two sexes were sometimes thought (incorrectly) to speak different languages. One of today's surviving Carib languages (Hixkaryana of the Amazon valley) surprised linguists by using the rare object-verb-subject as its standard word order. So in their language a sentence equivalent to "The boy hit the girl" turns out to mean the hitter was the girl and the boy was the hitee.

Despite this unusual structure, the relationships (subject and object) are familiar to speakers around the world. It was this ability to speak about the world in mutually perceptible ways that Columbus and the islanders discovered in one another. This observation seems so basic that we are tempted to dismiss it as self-evident, but if it is so inevitable, why don't apes learn to make sentences that combine a subject with an object? It seems implausible to suppose that apes cannot tell the difference between hitting and being hit. They know the difference. Shoichi did too, but knowing and expressing are two different things. Before our ancestors could express what, to some extent anyway, they already knew, they had to change dramatically without letting go of the speech triangle.

From this point in our history the tempo changes to reflect the constant interplay of biology and culture. The biological changes continue at a slow rate as genes quietly vary and spread, one generation at a time through a population. But now, there is a faster rhythm as well, reflecting cultural changes. Culture changes so fast today that people are inclined to forget about biology's role or even deny it altogether, but at any moment both the biological and cultural rhythms define a person's location in our history. The cultural tempo of *Homo erectus* was nothing like the madcap, sixteen-beats-to-the-bar pace of our present age. Some things, such as the *erectus* tool kit, remained stable for a million years, yet at the same time the species was pushing rapidly across Asia and onto some islands of the Indian Ocean. Classic fossils like Java Man (Indonesia) and Peking Man (China) show how far this ancestral species could travel and adapt to changing habitats.

The change in rhythm partly reflects the different means of proliferation. Genes spread the old fashioned way, one coupling at a time. Culture flies at the speed of sound from lip to lip, like butterflies pollinating a world of flowers. There is another reason that biology uses a slow metronome while culture skips on down the road. Biological changes must get the science right.

Evolving vision, for example, required developing nerve cells that responded to particular light frequencies, plus getting the light to focus sharply on these nerves, plus controlling the light's intensity so that the world looks like more than just a big bright spot. During the whole of this evolution, physical nature never helped out one bit. Light never adapted itself to the limits of the eye. Getting all that material precisely tuned to create a working whole took ample time, but once biology had it right, the job was largely finished. There may have been adjustments. Some animals are active all night and need to see well in the dark. Others have color vision, but for the most part the role of evolution became the highly conservative one of keeping the eye stable. We have changed in many ways since Sara's time six million years ago, but the insides of our eyeballs are still very much like a chimpanzee's. Adaptations to nature must work this way. Biological organisms do the adapting while fiercely preserving the details that function successfully.

Cultural adaptations work in just the opposite manner. Now it is the biological organism that stays the same while the cultural artifacts do the adapting. We repeat movie lines so that they suit our tastes rather than quote them exactly; we favor tools that fit our hands best, and we design our musical instruments to produce sounds we like to hear. We have already seen that the same principle applies to language. We organize our sentences so that we can follow them easily; we limit our vocabularies to what we can remember, and we base distinctions between sounds on differences that we can easily produce and hear. Thus, the story of how we got from single words to complex sentences has two separate beats. Cultural change accelerates; biological change is like Aesop's slow and steady tortoise. Cultural change is pushed by desire; biological change is not. But both shape the clay that emerges.

GALLERY II

Iskuhi | 1,799,925 Years Ago | Generation 350,005

A group of prehumans who are clearly the same species as Shoichi are stretched out along the shaded edge of a dry creek bed. They are immobile as they hide from the noonday sun. Behind them rises the bed's wall, climbing about two feet above their heads. The flat plain beyond the creek runs off to a mountain face in the far distance. The plain is covered with a bright light that makes the horizon ripple. About three hundred yards back stands a lone acacia tree where maybe fifteen wildebeest huddle together in its shade. The only visible creature that seems unconcerned with escaping the sun's heat is Iskuhi, a one-year-old toddler who sits up and pats the dirt. As she plays, she says, "sata."

Single-word utterances these days are associated with only very young speakers or angry bosses; naturally, we tend to think of loner words as tools fit only for children or childish behavior. Typical of that expectation is a scene like this diorama in which a small child plays by herself and says a word to which nobody gives the slightest notice. But if the first talkers were limited to one or maybe two words, we must suppose that eventually they put away the thoughts of a child and spoke as grown-ups. Iskuhi would never become as smart as us, but even limited to single words she was destined to outpace today's thirteen-month-old children.

The introduction of words was an even more profound environmental change than the move from forest to woodland. With words, our ancestors themselves began a long series of changes that challenge us to this day. Linguists sometimes compare the process to dam building by beavers or the corals that create their reef habitats. But while language does create its own environment, it differs from dams and reefs in the way that it has no end to its effects. Already by Iskuhi's time, life in a world of words was notably different from the pre-word life.

One immediate change was that individuals suddenly had names. Our tour gave names to the ancestors in Gallery I to individualize them, but now we are considering the lives of individuals who did have names, although we will never know what they were. That is something to remember when you read in the paper about a fossil discovery, some bit of old *Homo* jawbone and teeth from 1.7 million years ago. That jawbone was once an individual, known to friends by name.

Another new demand imposed by speech was the need to learn the words of the community and speak them distinctly enough to be understood. During the million and more years of bonding vocalization, the tone mattered much more than the articulation and there may not have been any compelling need to sound just like one another. With the onset of words, however, speakers had to sound much more alike. The need became especially pressing once the vocabulary expanded beyond a few words. The more words, the more important it is to distinguish between them. Eventually such precision of speech enabled speakers to distinguish between *cobble*, *hobble*, *bob-ble*, *bubble*, *babble* . . .

In this diorama, Iskuhi practices saying a word in her parent's language even though it has no immediate relevance to her situation. Besides talking during play, toddlers have been heard in bed at night practicing words.

Headphones provide a series of recordings—a zebra finch singing and the song of a humpback whale.

Although humans belong to the only talking lineage, we are not the only creatures who learn to imitate one another's sounds. A number of species learn to vocalize like their neighbors by listening and practicing. The words are not built into them from the start. Some whales and dolphins learn to make sounds like their neighbors, and some songbirds do as well. Notably, these other learn-by-imitation species, like humans, make combinatorial sounds; they build their vocalizations on a small set of sounds that they mix and repeat. If you take the sounds *er*, *form*, *in*, *mis*, and *t*, you can make *in-form*, *mis-in-form*, *mis-t*, *mis-t-er*, *in-er*, *t-in-t*, *form-er*, and *in-form-er*. Combinations like these are built in a novel way unlike the whole sounds that apes make. Ape cries have no syllable structure and use both the inhale and exhale portions of breathing to make their cries.

A pair of Dutch scientists, Bart de Boer and Willem Zuidema, have shown that animals who imitate sounds can start with whole signals like ape cries, but practice leads them to convert these unique sounds into strings of combinations. Mastering a sound bit by bit makes learning easier. Before mastering a complete sound, the apprentice vocalizers learn to make different parts accurately and finally manage to voice the sounds together.

Like Iskuhi, other creatures have also been observed practicing their sounds. In particular, scientists have studied how an Australian songbird, the zebra finch, learns to sing its song. A young male hears other males singing and begins to copy it. At first the copy is not very good, but finches remember the way a song should sound, and by comparing their output with what they remember, they slowly improve their performance. Eventually the birds can sing the song just as their neighbors do. Remarkably, the self-taught process depends on a specially modified gene that human speakers also use, the *FOXP2* gene. The gene also contributes to vocal learning in hummingbirds, parrots, and even some bats.

A model of the *FOXP2* molecule rotates on a steel pole.

FOXP2 is found throughout the biological world. All mammals have it, and they have almost identical forms of a gene that changes very little over long stretches of time. *FOXP2* controls the operation of many other genes, so there is not much room for tinkering. Usually a mutated *FOXP2* counts as a birth defect and is quickly removed from the gene pool. A rat and a chimpanzee have almost identical *FOXP2* genes; they differ by only one amino acid after 130 million years of separate development, but in humans the gene differs from the chimpanzee in two amino acids. The rate of selected mutations since our split with the chimpanzee line accelerated over forty times. Something important seems to have happened. *FOXP2* is also implicated in some grammatical constructions; however, that role would not become apparent until single-word speech was replaced by more complex sentences.

Many animals have a range of vocalizations that do not depend on the *FOXP2* gene for learning. A particularly interesting case is the domestic dog, which can express a rich variety of emotions by barking, whining, yelping, yipping, and growling. Wolves howl and utter a warning sound, but have nothing like the expressiveness of dogs. Perhaps this same kind of process occurred in the long preverbal vocalizing period of the human lineage before the *FOXP2* mutations.

Naturally, when *FOXP2* was first recognized, a media hullabaloo broke out over science's discovery of the "language gene," but *FOXP2* is not *the* gene. It is merely an important gene that contributes to the ability to speak distinctly. Today we have a list of over one hundred genes that respond differently to the human version of *FOXP2* than to the chimpanzee version, establishing as certain that the changed genes effects are numerous and varied. By now it is clear that the gene is critical to coordinating efficient vocalizing, but the fact that it has a role in such different sorts of animal vocalizing indicates that while it is a necessary gene, it is not sufficient unto itself to explain the biology of any species' vocalizations.

Originally, *FOXP2*'s mutations were dated at less than two hundred thousand years ago, making it younger than *Homo sapiens*. Many scholars were shocked by the recent date and thought something had to be wrong. Others, especially those who like the idea that language competence appeared in one big evolutionary bang, were pleased. Further study has shown that the early date must be wrong because the same

FOXP2 mutations have shown up in Neanderthals, a species of humans that separated from our lineage at least half a
million years ago. A reexamination by two researchers at the
University of Hawaii, Karl Diller and Rebecca Cann, put the
date much more in line with the rest of the evidence, dating
the *FOXP2* mutations from 1.8 to 1.9 million years ago, a bit
before Iskuhi's time.

It may seem like a miracle that the modified *FOXP2* came
along just when it was needed, but do not forget about selection. Every individual is born with mutant genes, but selection
typically works against them, especially in super-conservative
cases like *FOXP2*. In Iskuhi's day the mutated gene and its
importance to vocal learning was suddenly accepted as just the
thing to give Iskuhi's group an advantage over rivals.

> Iskuhi, at about age six, and another juvenile of about the same age
> are squatting, tapping the ground, and chanting together, *"ba-bee-
> boo-boh..."*

Playing with the sounds of speech is a source of joy, and,
like music, it is an affront to all the serious thinkers who want
to remove happiness, sensation, and whimsy from the list of
primary human traits. These elements are all irrational, and
an important school of thought about human behavior focuses
exclusively on rationality—true/false propositions, computation, and personal interests. This history, however, belongs to
a broader tradition and assumes that vocal play has been with
us from the beginning.

By Iskuhi's time, vocalizations already had about 1.4 million years behind them, and bonding by vocalizing did not disappear with the onset of words. Words, however, changed the

sound system they used, making it dependent on syllables. An interesting objection to this history is that chimpanzees and other apes have a difficult time making the vocalizations of language. One family that raised a chimpanzee from birth as though it was their own child found that despite their hard work the chimpanzee only managed to say one word, *cup*. The first speakers very likely had the same vocal tract as a modern chimpanzee, so how did they manage to use syllables, playfully or not? The chief problem facing chimp trainers, however, is that chimpanzees have no motivation to try and make a human-like sound. Physical calculations indicate that a chimpanzee's vocal tract is capable of making as many as ten different morphemes, the basic units of meaningful speech sounds. It is not many, but enough to get started.

Our own vocal tract is more flexible than Iskuhi's, so there must have been some biological evolution as well as a cultural one. The tempo of this co-evolution will have been the same as the tempo of language evolution in general. The cultural change runs; the biological change walks. Only when the culture has begun using the available sounds will there be any biological pressure to enlarge the repertoire. At that point, the groups compete over repertoire size with the ones able to handle the richest culture outlasting the others. By now, the range of possible variations is great and languages have distinct sounds. In the early days of speech, there may have been much less variety in the ways different groups sounded.

Iskuhi, still about six, is with the same friend in the previous diorama. Both are looking down along the slope of a volcanic crater. Far below is the crater's grassy floor. Halfway up the crater wall stand two grown zebras and one foal. The friend points at a large zebra and Iskuhi says, without pointing, "*Tother.*" [Obviously these youngsters did not speak English words like *tother,* but there is no point in inventing a series of languages just for this history.]

Even with single words and even at a young age, we can be confident that speech went beyond the level of a baby's words. Genetic factors matter little in shaping vocabulary. Words are primarily dependent on cultural sources, yet even here biology plays a role. Apes can probably perceive and keep track of as many as three separate things at once—perceiving *this, that,* and *t'other.* Many languages have ways of identifying three variables individually. The words they use are culturally defined, contingent upon some specific history, but they do not evolve randomly. They follow a path that reflects their perceptions. We take for granted the ability to follow three separate things in a sentence—e.g., *Joe told Pete about Sally's wedding*—and we do not wonder about the biology that supports it, but the powers of perception have their own long history and those powers secretly shape languages, including the words speakers use. We identify variables by *it, this, that,* and *t'other,* because we can keep track of four things at once. We don't add as variables *miflix, miflon, mappenflor,* and *maxenfeffer* because most of us cannot keep track of all those separate things at once—*Joe told Pete, Bob, Carol, Ted, Alice, Henry, Janet, Larry, Kathy, and Augie about Sally's wedding, Charles's divorce, Myrtle's tenth-anniversary party, and Arbuthnot's hope to find true love in Memphis.* Some linguists would accept this monstrosity as a sentence, but no editor would. To put it in twenty-first-century terms, speech that survives is user-friendly.

Iskuhi, now a half-grown juvenile, is with her mother looking across a flat, grassy plain. Off to the left, perhaps twenty feet away, are some others, and far out on the flatland by himself is a lone male. Iskuhi and mother are both staring at the solitary male. Iskuhi says, *"Heyoka,"* naming the loner. Her mother, in a strong tone, stresses a different name, *"Yuhamani."*

Differences in perception are as old as perception itself, but only with the introduction of words could anyone learn that two individuals standing side by side might disagree about what they see. Probably the speakers of the early talking generations like Iskuhi's, when speech was still very simple, found these perceptual differences impossible to understand. Iskuhi cannot say, "Oh, I was wrong." And the mother could not say, "There is a resemblance, but if you look closely you can see who it is." Instead, the most that seems possible is for Iskuhi to follow her mother's direction and see Yuhamani.

Such transformations of identity sound much more dream-like than rational. Iskuhi lived long before the age of mythology, but she and her generation had already moved into a world where things and people changed their natures with magical ease. For scores of millions of years, Iskuhi's ancestors had accepted their perceptions without question, but speech complicated matters. Iskuhi lives in a world where a name could be useful, and then not.

Three near-grown juveniles press their backs against the rocks of a koppie. A fourth—Iskuhi—peers over one of the rocks and says, *"go."*

Despite its simplicity, this scene illustrates one of the most important novelties of language, the need for speaker and

listener to hold a common mental ground where they can focus their joint attention. Iskuhi's companions know what she
is talking about because they jointly attend a shared mental
ground. We, however, are outside that ground and do not
know. Perhaps a lion is going away from the koppie. Or maybe
an angry adult has gone. We cannot know because we do not
know the topic that binds the group together. Even with speech
limited to single-word utterances, the introduction of topics
permanently changed social relationships, dividing groups into
those who know the topic and those who do not.

Before the appearance of words no such division existed.
An ape or an elephant or a bird could make its sound and
no matter how complicated it was, others in the group got
the message and could respond appropriately. Now, along
with the ability to recognize the utterance, the listener needs
to know the topic. Perhaps nothing illustrates the uniqueness
of language more than this dependence on a common ground.
The dance of the honeybee provides unusually specific information about how to find pollen, but locating pollen is always
the point of the bee's dance. There are no insider and outsider
bees. Language's capacity to turn anything into a topic allows
for a tighter bond than any previous groups felt, but it can also
create a feeling of exclusion that is probably more profound
than anything known by a chimpanzee.

The need for a common ground is especially apparent
on this koppie because only Iskuhi can see what she reports.
The others must remember what she is talking about. Perhaps
the others have never even seen the topic. Iskuhi might have
looked over a rock and said, "*lion*." Each of the others then
individually remembered the image of the most ferocious lion

they ever saw and pressed against the rocks. Besides directing attention to something present, words can direct attention to absent things that must be imagined. Linguists call this ability to talk about absent topics "displacement" and often consider it among language's most remarkable powers.

At least as radical is the way language—even at this one-word stage—has pushed part of life directly up to the conscious level. Awareness, of course, is very old. Attention depends on it, but much of animal life is passed as reflex and habit. Animals come to attention only from time to time, studying a scene and acting on the basis of what they find, but language works by directing attention and thereby forcing topics into awareness. Those juveniles pressed against the koppie are conscious of the danger that is nearby. If it is a lion they dread, the word *lion* and the associated image work together to keep the thought locked into consciousness.

Also, language is always ambiguous. Some confusion can be cleared away by sharing common ground with a speaker, but other material is inescapably vague. This problem is visible at the koppie with Iskuhi's use of the word *go* rather than *going* or *gone*. Most accounts of language evolution assume that the first utterances lacked markers to change verbs and nouns. The work of a Serbian American linguist Ljiljana Progovac has been especially valuable for its study of primitive, or pre-syntactic, clauses, and she notes that such clauses use a consistent form. In English the default verb form is the infinitive. At this early stage in speech origins, a word like *go* could be used in a situation where modern speakers would mark for tense (e.g., *gone*, *going*). At the dawn of speech, selecting an interpretation probably depended on what Iskuhi did after she finished speaking. If she started running, she probably meant *let's go*. If she went forward over the rocks, she likely meant *it's gone*. If she stayed pressed against the rocks, maybe she meant *it's going*. Although language will eventually become more informative and able to sort out some of these ambiguities, it has never become so transparent that common ground and non-verbal cues could be abandoned. Probably, language evolution would actively resist such transparency. Part of the point of group cooperation is keeping outsiders out, and a perfectly intelligible language would let strangers in.

> Iskuhi stands with a child of her own. On the horizon are three giraffes. The child looks toward them and asks Iskuhi, "*What?*"

Up till now we have seen declarative talking that identifies a topic, but does not necessarily invite a dialog. Asking questions, however, demands vocal interaction. A child asking a question simply assumes many elements of a speech triangle—knowledge of a topic, joint attention, and the ability of speakers and listeners to change roles.

> Iskuhi sits beside some rocks. She is making a cutting tool in precisely the same way that Ng'ula did three hundred thousand years earlier.

The two lines of evidence about *Homo* culture in Iskuhi's time contradict one another. The fossil evidence shows a brain that got steadily larger. E. O. Wilson has said that for two and a half million years our brains grew by about a tablespoon per hundred thousand years, which works out to about two-thirds of a cup per million years. Meanwhile, archaeological data compiled by scholars unearthing tools and other material remains of early *Homo* is remarkably stable. Oldowan technology, as these first intermediary tools are known, stayed the same for about nine hundred thousand years, suggesting a serious intellectual plateau. Different sites show that different kinds of stone were used and there were also different techniques for flaking off the cutting blades, but the end result was remarkably consistent over scores of thousands of generations. A recent analysis of these Oldowan tools separated by time and geography concluded that they all reflected one toolmaking tradition. Fossil troves in the southern Caucasus of southwest Asia show that by 1.8 million years ago, *Homo* had taken the toolmaking tradition beyond Africa. Many Oldowan tools survive from that epoch, buried in the mountains of southwest Asia.

How are we to resolve the tension between the fossil evidence of growing intelligence and the archaeological evidence of no increase in cleverness?

First, by remembering that diversity and innovation come mainly from already existing diversity and innovation. Evolution—be it biological or cultural—builds on what is, and in the beginning there was very little. Biological evolution gets no help at all from intelligence, and for most of the history of life—starting about 3.5 billion years ago and persisting until almost six hundred million years ago—it got no further than pond scum. Then multicellular life appeared and the pace quickened.

Cultural evolution does have a place for intelligence, yet it still builds on what already exists. In its early days, there were not many items to evolve, and anything novel that did appear had to compete with many more instances of older items. We still see this stabilizing phenomenon even in our own day, as in the case of irregular words, mainly verbs. The standard rule for expressing an English verb in the past tense is to add -*ed* to the end (*create*, *created*), but there are many exceptions (*run*, *ran*; *eat*, *ate*; *swim*, *swam*) and these exceptions are common words whose very frequency of use preserves the knowledge of their irregular form. The more frequently a word is used, the slower the process of taking on a modernized form. In the early days of language there would have been only a few words, and they got steady use. Big changes were likely to be rare events. There was a kind of heavy gravity for resisting any change. Yet subtle changes occurred. The brain evidence must count for something.

A photograph of a page from the Oxford English Dictionary shows the word "alternate" and its thirteen definitions. The original meaning dates to 1513, but variant uses kept appearing (1590, 1650, 1660, 1697, 1712, 1716, 1718, 1809, 1848, 1858 (twice), 1861, 1898, 1961).

A tradition as fixed as the Oldowan is unlike anything modern humans can imagine from their own knowledge of technological history. No human society today is anywhere close to being so conservative that it uses the same tools that were used nearly a million years ago. Presented with evidence of such a hyperconservative culture, we are forced to wonder if its speech was equally stable. Could language have kept the same words for a hundred thousand years or more? Maybe even a million years?

There is a strong theoretical argument against such linguistic stability. If a word persisted for a hundred thousand years, it might be so stable that evolution would build the word into the brain. Then language would have two kinds of words:

inbuilt, universal ones, and cultural ones that are passed down from generation to generation. But we do not have such a dual word system.

Or do we? Remarkably, it is possible that a word from Iskuhi's time has survived even to this day. That word is *mama*, a word for mother in English, Swahili, Samoan, and many other languages. In fact, the word is so common that some people have proposed that it traces to some unusual, non-linguistic origin such as the sucking motion of the lips made while feeding. Could that be an inbuilt word? Closer examination, however, does find a few telling exceptions. On the island of Tonga in the South Pacific, the first *m* has become *b*, *bama*. In Albania, the first consonant has disappeared altogether, *ama*. Sanskrit used *nana*, a word for grandmother in some other languages. And a small Australian group, the Namal, does say *mama*, only it means father.

A team of French-led scholars examined the few words that are found in 60 percent or more of the world's languages—kinship terms like *papa*, *mama*, *nana*, and *kaka* (mother's brother)—and found that, as we saw with *mama*, the words are not perfectly stable. For example, the Turkic word for *papa* is *apa* and is the same in most of its descendant languages. (Turkic was the language of a nomadic people of central Asia, and its descendants have spread very far, from eastern Europe to central Asia and northern Siberia.) This consistency through descendants is exactly what happens with any other common word in a language, although in this case the change is geologically slower.

The team also considered where the pressure for such stability comes from, and pointed to the earliness of the words' appearances in language. A mother, father, granny, and uncle are among the first people known to almost any child, and they are identified while a child's speech is limited to the basic babble sounds. Any words a child uses for these people will inevitably be composed of simple little syllables.

Does this mean we can say with any kind of confidence that *mama* was indeed one of the very first words to appear in the rise of speech? Sadly, no. "Sadly" because it would be satisfying if we could say that we knew at least one word used in deep time. "No" because there is an impenetrable wall of ignorance concerning the details of speech older than humanity's last common language. The date for that unknown tongue can be no more remote than 170 thousand years ago, when humanity's last common ancestor lived. And it may have been much more recent. A paradoxical finding puts the last common male ancestor 110 thousand years after the last common female ancestor, who lived only sixty thousand years ago. The paradox comes from tracing common female ancestors according to one genetic feature (mitochondria) and male ancestors by another (Y chromosomes). Common sense tells us that if we are all descended from a Mister X, we must also all be descended from Mister X's mother, who did not live 110 thousand years before her son. Thus, the evidence is merely contradictory, not paradoxical, with the weight of the evidence leaning more toward the female date 170 thousand years ago. It seems very unlikely that northern European and southern African peoples share a common ancestor as recent as sixty thousand years ago. In either case, at least 90 percent of speech's history was already past. Thus, just because speakers of the last common language were almost certainly calling their mothers *mama*, that's no evidence that their ancestors over one and a half million years earlier used the same word.

More important, however, is that even during the early, hyper-stable period of *Homo* technology, language was being passed on in the same way it is now, from speaker to listener. There is no set of words that were built into the brain, either as a starter set of sounds or from some form of biological evolution.

Why not? If culture was so stable that tools stayed the same over a period of hundreds of thousands of years, why weren't evolutionary processes strong enough to hard-wire at least a few early words? The most obvious answer is that speech was much less stable than toolmaking, and in fact we have evidence that this answer still holds true today. Modern dwellers along the Nile can be seen using tools like sickles, knives, and even lifting devices that are depicted in ancient Egyptian paintings, although we know for a fact that the modern people speak a different language from the ancients.

It is also true that toolmaking is passed along in a different way than speech. Masters teach apprentices how to make tools the standard way, but once the lessons end an apprentice is able to act alone. Speakers and listeners, by contrast, interact continuously over the whole of their lifetimes.

> A troop of ten, including Iskuhi, who is now a young adult, are scattered across a plain. Against the far horizon is a volcano in full eruption. All the prehumans are looking at it, and the Thomson's gazelles have come to attention as well. Iskuhi stares at the mountain and says, "Wow."

Emotional expressions are much older than speech, so Iskuhi seems to be simply voicing an animal signal, but animals do not say *wow* or anything of the kind. At most, they might make a cry of fear, but that is a product of unconscious reflex and encourages no further notice. *Wow* sounds like an automatic cry, but it is part of a language. Other languages use other words to express wonder and amazement. Like other words, *wow* reports news about something, in this case the news is the response itself. Once again, language is forcing something into consciousness. Ancient emotions that have gone unnoticed or have been immediately forgotten have become a part of life.

Iskuhi stands making rhythmic thumps on her stomach while saying *"trala."* Two children, one about four and the older nearly an adult, join in the singing.

Singing can be a bridge to proper speech. The person with a marked stammer may be able to sing without giving a sign of a problem. The rhythms and reliability of the audible pattern seem to help, and when trying to learn a new multi-syllable word, especially in a foreign language, people can rhythmically sound it out and speak it that way. For children and newcomers to a language, the brain has not yet absorbed the system of sounds specific to a language and cannot yet produce them with a reliable, easy fluency. Rhythm and prosody lets the mind stretch itself, accomplishing a bit more than normal, like a person with Parkinson's disease who shuffles across a street and then looks lively when prodded by an oncoming honking car.

Iskuhi has grown to be an older female. She is standing guard while a five-year-old boy drinks from a stream. Far against the horizon is the volcano, quiet for many years now. Iskuhi whispers, *"Albert,"* the name of a long-dead youngster.

For millions of years, animals have been briefly reminded of outdated associations. It is an inevitable part of learning and perception, but normally the reminder promptly fades because the animal can make no use of it. Without words, animals have no way to grasp the association. With the appearance of names, however, there is a double association. The act of guarding one child by a stream reminds Iskuhi of the same act of guarding another child, but now she has something to associate with that reminder, a name. And the name forces the lost child back into her consciousness, perhaps with still other associations. Unlike the displacement at the koppie in which a word referred to an invisible space, this displacement refers to a lost time.

All of these changes shown by Iskuhi—combinatorial sounds, displacement, consciousness of absent things, expressions of wonder, and a common mental ground with others—were immensely liberating. Sound combinations removed the limit on the number of possible topics to consider. Displacement of time and space ended the boundaries formed by the here and now. Consciousness allowed her to examine a topic, even a displaced one, more carefully and let her make use of still further named associations. And finally, a common mental ground enabled others to learn about all those shared topics. Even at the single-word level, the presence of speech radically altered the experience of life.

> An aged Iskuhi and a small toddler huddle together in the rain. They stand silently, looking toward a patch of brush a hundred yards away where a pair of lions have taken shelter from the storm.

A reminder: Despite Iskuhi's one-word speech and high intelligence, she was far from being among the top predators on the plains. Life remained dangerously prone to death by misadventure, and even the lucky found the days harrowing.

Gärd | 1,799,870 Years Ago | Generation 350,010

Gärd and three children are by a shallow stream. The two younger ones are playing together while the oldest is drinking from the stream with cupped hands. Gärd is looking off in the distance and says, "*hyena*," and then adds, "*big*."

The leap from speaking lone words to phrases may have taken place within five or even fewer generations. After all, no biological changes were needed to find the intelligence to put words together. Apes today trained in signing will sometimes spontaneously create signs for phrases. Probably, however, the first spoken phrases were not terribly fluent. In this diorama, Gärd has spoken two words together. More importantly, the second word modifies the first. She has not randomly associated two words but has voiced a phrase, specifying something and describing it all at once. The value of adding the word *big* is probably that it makes the urgency of the situation more palpable. *Big* does not connote just large, but dangerous as well. With such an impoverished language, Gärd is in no position to give a series of instructions like "Gather stones, stand together, protect the baby." The family will have to act cooperatively without much instruction, but the news that the mother has perceived a dangerous animal stresses that it is time to act.

Using a word to modify another marks an important change from single words. It signifies, at a minimum, a richer attention to a point of focus. The regular use of two-word utterances in children surely indicates some mental growth. Neurologically, speaking phrases requires developing pathways in the brain that make it possible to combine words together in a stream of sound, even if it is a hesitant stream. Gärd has the intelligence to use one word and then another. A gain in fluency is a natural reaction, the creation of a brain circuit that makes a repeated task faster. The better a task is learned, the more quickly it can be performed and the less brain activity it requires. Developing appropriate neural pathways allows an individual to accomplish more without using as many calories. Learning pays for itself by making an action more efficient.

The development of a more efficient brain was probably one of the reasons the human lineage was able to grow such a large brain. Already by Gärd's time, the brain was twice the size of an ape's. A number of factors must have contributed to this steady growth. We saw that there was an immediate gain when cooperative communities appeared. That pattern suggests that as speech made cooperation even richer, the community could support ever larger brains. Speech is probably not a luxury, but something that paid for itself. Harvard anthropologist Richard Wrangham proposes that by Gärd's generation, *Homo erectus* was cooking food, which provided many more calories to support the brain without increasing the amount of food gathered. It is an important idea, one with decent supporting circumstantial evidence; however, there is no physical evidence of controlled fire until about 790 thousand years ago, almost a million years after Gärd's time.

A circle of arrows are shown on a wall poster. At the poster's top is the phrase "Enlarged brain." An arrow arcs down to point from the top phrase to a second phrase, "Better community cooperation." Another arrow arcs down to the poster's bottom phrase, "Brains become more efficient." A third arrow arcs back up to the words "Efficiency allows brain growth." Yet another arrow completes the circle by arcing back up to the words "Enlarged brain." The circle goes round and round.

This history assumes that the growth of the brain was a response to living conditions. The general process was circular: *Homo* communities were under selective pressure to speak well, and the enlarged brain made it possible to learn to speak more richly. This enriched speech led to more varied community cooperation; supporting this enrichment, the brain evolved structures that made the new speech even easier to learn. With this newly efficient brain, the species could afford to support a still larger brain without requiring a commensurate increase in calories . . . taking us right back to the beginning: The enlarged brain made it possible to learn to speak even more richly.

Behind this circular process is the steady need to adapt to life in a community. Each adaptation has new consequences requiring further adaptation. Naming leads to a pressure to be more specific which leads to . . . Well, we shall see as we tour our exhibition, but the pressure to cooperate was relentless and drove the process around and around in a circle of learning, more efficient action, and brain enlargement.

This kind of goal-driven learning is familiar to anyone acquainted with today's world. Schoolchildren see a classmate master a feat, perhaps jumping two twirling ropes at the same time, and they practice what they saw until they can do it too. Artists can be dissatisfied with a work and invent some new technique that lets them do what they could not do before.

Scientists may realize that to solve a problem they will need to master some obscure bit of knowledge—as, for example, Einstein had to learn non-Euclidean geometry in order to create his general theory of relativity—and they go about teaching themselves the subject. Goal-driven learning is less common in the animal world, although anyone who has ever watched a bear break into a garbage can in Yellowstone National Park knows that animals too can sometimes teach themselves what they need to know.

The dissatisfaction a *Homo erectus* like Gärd felt with single-word expression may have been similar to the dissatisfaction a poet feels today with a first draft. "Hyena" by itself does not express what she wants her young to know. We do not normally think of ancestors from almost 1.8 million years ago as having poetic urges, but that prejudice comes from our notion that poets are ivory-tower eccentrics. Dissatisfaction with the limits of speech, however, is quite common and likely to be almost as old as speech itself.

Beside the poster showing the circular process of brain growth is a photograph of James Mark Baldwin. A bit of Princeton University's old library is out of focus in the background.

The process by which something is first learned and then becomes part of the genome is known as Baldwinian evolution, named after James M. Baldwin, an American psychologist who described a Darwinian way to reverse the more familiar process of altering a gene and then selecting the result because of its greater fitness.

In basic Darwinian evolution, for example, a mutation (or to use Darwin's term, a "variation") might make the control of the tongue more precise so that some individuals are able to make their sounds more consistently; over time, the variation is selected and becomes a universal part of the genetic inheritance. The classic Darwinian pattern is (1) variation, (2) change in the resulting individual, and (3) selection of the variation until it becomes normal in the species.

With Baldwinian evolution, the change comes before the variation. This kind of change reminds biologists of the pre-Darwinian theory in which Lamarck proposed that, say, giraffes first began to stretch their necks and then the offspring were born with longer necks. Lamarck, however, did not include variation and selection anywhere in his account, and he made the process the universal explanation for the descent of species.

Humans, with so much culture and learning, are particularly subject to Baldwinian evolution. An example that has gained much attention in recent years is the evolution of the ability of adults to digest lactose in milk. Around the world, some cultures have become pastoral—herding cattle, goats, sheep, and camels and living off the produce of the herds. Over time, the herders became genetically able to digest the lactose in milk of the animals they herded. That adaptation is a clear non-linguistic case of behavior-first Baldwinian evolution. Much of human evolution reflects this Baldwinian shape: A cultural change is later supported by a biological change.

There is a catch, however, when Baldwinian evolution is applied to speech: Language changes faster than the genome. Thus, the only parts of language that Baldwinian processes can build into the brain are the universal ones. Exactly what constitutes language's universal elements is a hotly disputed matter, but there are some things that everyone agrees on. For example, all languages string words together. Thus, the ability to combine one word, e.g., "hyena," with another, e.g., "big," is exactly the sort of thing that might be built into the brain. Be sure to notice that it is the *ability* to combine that is universal, not particular combinations. In the diorama, Gärd pauses between words. The ability to string them together still more fluently will appear later.

The evidence for a three-step process in which fluency comes last is strong, but circumstantial. People who suffer strokes often lose some language ability. At first the damage to their brain may leave them mute and uncomprehending. The course of recovery depends on the extent and location of neural damage, but in general the path to recovery begins with comprehension and then a recovered ability to speak correct sentences. Fluency returns last. Similarly, a person learning a new language first speaks individual words, then hesitant word combinations, and finally new phrases. This account of speech origins assumes the same pattern occurred originally.

> Gärd and two others look toward a fever tree with three impala draped over a yellow branch. No leopard is in sight and Gärd says with ease, "*gone far.*"

An important point to keep in mind as we explore the origins of speech in this pre-sentence period is that even with two-word phrases, language could not yet combine a topic with news. The best speakers could do was voice a single phrase that brings news about an understood topic. The Russian psychologist Lev Vygotsky maintained that thinking (what he called "inner speech") consisted of mostly news (known to grammarians as a predicate) with an understood topic (or subject).

A predicate is news about a topic that directs attention away from the topic itself. In this scene, *gone far* is a predicate. Presumably a complete sentence would be something like "That old leopard has gone far." In Gärd's utterance, the "that old leopard" part is understood. It must be understood since there is no point in reporting news alone unless the topic is already part of the speaker and listener's common ground. Yet understanding and sharing a common ground is not the same as speaking something. Gärd still directs attention to only one thing. Even if she has become more fluent at two-word speech, she is still not thinking beyond the ape level. However, Gärd has taken an ape's linguistic potential much closer to its maximum.

Individuals were not yet speaking in full sentences and could only refer to one side of an idea: either a topic or the news, the subject or the predicate. Linking a subject and predicate is so natural and easy for us that it is difficult to imagine what it would be like to be unable to form such a link, or to form it only with great effort. But this kind of half idea is probably what dominated the human lineage of one and three-quarter million years ago. Still, *Homo erectus* could do more with a phrase than a two-year-old *sapiens* can.

> From a distance, Gärd and three others look at a fever tree where a leopard has hung a gazelle. Two lions stand directly below the gazelle. They have been drawn by the ripe odor but cannot recognize the animal as it dangles upside down above them.

With their bigger brains, Gärd and company had no trouble recognizing what lions could not. Their advantage over the top predator is apparent, although whether they are up to robbing a leopard's cache is another matter.

An adult male is sitting beside rocks, chipping a cutting flake from a core stone. Gärd looks on and says, "*want.*" The toolmaker ignores her. Gärd says, "*want it.*"

Chimpanzees are smart enough to imagine a desirable outcome, and to change their social behavior when that outcome does not look promising. Thus, we have ample reason to suppose that Gärd, with twice a chimp's brainpower, could use another word if the first did not suffice. Again Gärd has almost expressed a full idea. If she said *I want it*, most of us would not think it much of a breakthrough, but that phrase combines two points of attention in one statement. Here she focuses on just the news and leaves the topic to be understood.

Gärd looks at the toolmaker and sings, "*waaaaa-aaaaaant iiiiiiiiiiiiiit,*" making each *i* lower in pitch than the last.

Neurologically, there are two processes here: the production of the sound wave by the vocal cords and the articulation of the sound into words by the vocal tract. The two are controlled by different systems in the brain and speaking requires synchronizing the two. They can, however, be deliberately separated, permitting the extended vowels of the playful person, the insistent one, and the opera singer.

Gärd says "*hear heart*" three times, fluently but not accurately. The first comes out right, "*h-eer h-art,*" but the second contains a stumble, *"h-eer h-eert,"* corrected to *"h-art."* The third time is a failure as *"h-ar h-art"* is corrected to *"h-eert h-eert."*

Playing with words, and even tongue twisters, stretches vocal skills. Anne Karpf writes, "With the exception of the muscles around the eyes, those of the human larynx have more nerves than any other muscles in the human body, including the hand and face, even though we only use around one-third of their capacity in speaking. Each can produce a different balance of forces in the larynx, generating a different pulse wave and sound quality. They're our vocal palette: through them we color our voices with affection, bitterness, pleasure, disgust." Such details remind us of the sheer physical development that was required to make us modern speakers. We have much more accurate control over the sounds we make than apes have, and we also have a wider range of sounds than apes. This kind of control and range has been developing side by side over the ages with the expansion of vocabulary and rise of syntax.

Gärd and three other grown males are standing together on a grassy plain. The horizon is just a skyline with no trees anywhere in sight. In the distance is a topi antelope with dark, reddish-brown skin. About ten feet away from the foursome is another adult male, holding the meaty leg of a Thomson's gazelle. Speaking fluently, Gärd says in an accusatory tone of voice, *"take bones."*

The compound *take bones* is not known historically, but it is the kind of verb-noun idiom investigated by the Serbian linguist mentioned earlier, Ljiljana Progovac. She looks at word combinations that neither link a subject to predicate nor do they modify one as with a noun and an adjective or a verb and an adverb. Progovac sometimes refers to these bits of language as "fossils" because they reflect a structure of language that predates the formation of modern syntax. Of course, something in use today cannot really be a fossil, but the forms are syntactically primitive.

A modern sentence consists of phrases that combine to become a subject and a predicate. The words in a phrase modify (or shape) one another. Listeners can understand a phrase by combining the meaning of its words. Even a person who speaks only French can decipher an English phrase by looking up its words in an English-French dictionary. *Greedy Henry took all the bones*, for example, contains two phrases, *greedy Henry* and *all the bones*. We call both of them noun phrases because, like nouns, they identify things. In formal linguistics, *took all the bones* is called a verb phrase because, like a verb, it defines an action. It breaks down into a verb and a noun phrase.

Now look at the phrase *take bones* and note how it defies this formal classification system. First, neither word in the phrase modifies the other. Real examples of this kind of combination include *turncoat*, *daredevil*, and *tattletale*. *Turn* does not modify *coat*, nor does *dare* modify *devil*, nor *tattle tale*.

A second deviation from formal rules is that while *take bones* looks like a verb phrase with its verb + noun structure, it works like a noun phrase. *Take bones* names a thing.

Also, the meaning of the combination cannot be determined by looking the words up in a dictionary. A *turncoat*, for example, is a traitor, although you cannot figure that meaning out from the word itself. The same holds for *take bones*; its meaning is undecipherable from the words themselves; even so, we might divine that the term refers to people who eat every scrap of food by themselves without sharing. A modern speaker might say, "You are a selfish pig," but someone could say simply, "selfish pig," and the *you are* part would be understood. *Selfish pig*, however, is a modern phrase with a modifier. *Take bones* is not.

We saw Iskuhi change a perception when corrected by her mother. A compound like *take bones* also changes a listener's perception by asserting a new identity for a comrade of many years. The change is magical, not logical and not syntactical.

> The same scene as above, a moment later. The *take bones* looks at Gärd and says with equal fluency, "*curl lip.*"

Here we have another verb-noun compound of the type that Progovac describes. In this context it appears to be the equivalent of "liar," bringing home a sad conundrum about language. When speakers disagree, whom do you believe? In this case, the *take bones* appears to have been caught holding the equivalent of a smoking gun, but with the introduction of language comes the problem of authority that has bedeviled the human race ever since. Talk can be invaluable, but it is also cheap.

> Gärd and seven other adults are throwing stones at *take bones,* who is fleeing into the open savanna. All eight of the stone throwers are chanting, "*take bones … take bones … take bones.*"

Admirable as group-level selection may sound, it is every bit as brutal as selection at the gene level, and some of its brutality can be deliberate. Anyone who fails to live up to the group's cooperative expectations may be subject to fatal penalties such as exile. A lone *Homo* on the open grassland would be lucky to survive two nights. *Take bones* is running toward his death.

This kind of group punishment for selfishness is a rough solution to the problem of the selfish gene. If *take bones* is not punished for his selfishness, any genes supporting that selfishness can spread and eventually undermine loyalty to the group. At the same time, the group is not an anthill. The members are individuals with different abilities and degrees of authority. The tension between ego and group is as old as the *Homo* group itself.

> Gärd and the seven other stone throwers stand together chanting sounds that create an emotional sense of union. They are not chanting words, merely sounds, but that simplicity is a strength because there are no distracting shifts of attention. The chant keeps the emotion focused on the solidarity of the chanters.

Dakila | 1,799,805 Years Ago | Generation 350,015

> Dakila shows two near-grown juveniles how to make an Acheulean axe. *"Like me,"* he says and hammers on one side of the stone. The juveniles watch closely.

The Acheulean axe was a new tool in Dakila's time and was then passed down through thousands of generations with limited verbal instructions. There is some dispute over how readily the next generation could learn the craft of making such an axe, especially in the early years. Dakila was only fifteen or so generations removed from the first talkers. Linguistically and intellectually, he was probably still a scaled-up ape. Socially, however, he was a kind of human being. The first *Homo* was already a million years dead, plenty of time for community-focused urges to become deeply fixed in the population. One unusual community-oriented trait found in humans is imitation.

Included on Judy Shepard-Kegl's list of foundational traits required to support the creation of a language is "a taste for mirroring." Seeing what others do, say, or gesture leads to imitation and adds to the common pool of words and actions shared by a group.

Faithful imitation is devalued in our time and creativity favored, but pragmatically there is much to be said for copying. Imitation can be mindless, but with experience the imitator can come to understand the value of an action originally learned through blind imitation, advancing from being a copier to a participant in a culture. In the case of axe-making, the juveniles may not yet appreciate the biface tool, but Dakila can help them to see that the two sides come together by telling them what to do.

A taste for imitation is a taste for ratifying the speech of others. It has often been suggested that the language instinct is for grammatical forms, but at least in the early stages it is an instinct to imitate and use what others say.

Yet even with an urge to mimic another's actions, making a biface tool, like speaking a complex sentence, requires more than learning associations. You have to be able to link sets of associations together. We do it so easily that we do not give the fact any special thought. Actions that require organizing sets of associations are based on "sequential learning." An example of a challenge requiring sequential learning that is faced by many primates is obtaining fruit that cannot just be picked from a limb. The picker may first need to remove thorns and then pull aside a barrier of tightly packed leaves. Experiments have shown that chimpanzees can learn an arbitrary four-action fixed sequence. So it seems reasonable to suppose that the big-brained *Homo* of 1.8 million years ago required no further evolution of the brain to learn the procedure for working both sides of a stone before making an axe blade.

Even so, the introduction of regular toolmaking through sequential learning may have helped with speech. A psychologist and investigator of language origins, Morten Christiansen, reports that stroke victims who have severe problems organizing sentences also have trouble with sequential learning. Even more striking, training stroke patients in complex sequential learning that does not involve language leads to "improvement" in complex language production. So there is good reason to suppose that generations of axe-making would, through changes in the brain, lead to more complex speech.

If that change sounds magical, remember that the great unchallenged work of Noam Chomsky and his school of linguistics has been to establish that sentences are not just strings of words placed one after the other like markers on a road map. They are phrases—sets of associations—joined together. Take, for example, the sentence *The two leaning trees can be seen from the spring.* The standard way of parsing the sentence is to say it consists of a noun phrase (the two leaning trees) and a verb phrase (can be seen from the spring). The verb phrase itself is formed by a verb (can be seen) and another noun phrase (from the spring). A functional analysis of the sentence is a little different. The two points of attention (the noun phrases) are made coherent by the verb. The verb itself makes sense with either point of attention, but if you try joining the two points without the verb the result is confusion. Both analyses are hierarchical, combining sets of associations into a string.

However, we are jumping ahead of ourselves. Dakila's time seems too early to start linking sequences of associations. *Homo* had just begun understanding a sequentially based technology. Still, sequential learning was probably not confined to axes.

Dakila and another male stand beside a narrow pool of water on the open grassland. They hold their hands together and raise them above their heads while both, more or less together, say, "*us*."

Here is another example of how two separate things (gesture plus word) can be combined into something new, in this case a ceremonially based relationship. Ceremonies that ensure trust and a serious sense of obligation are common in human affairs, although they are unknown in the rest of the animal kingdom. Animals do have ritualized behavior in which they establish dominance, claim territory, or win a mate, but these actions all concern individual biological needs. Humans have them too, but we also have many ceremonies that address more communal needs. Ceremonies do not have to be public, like weddings and funerals, and many are small and simple. Shaking hands to conclude an agreement is a classic example of a private ceremony. The ceremony grounds an agreement in a physicality that would otherwise be absent.

Even simple ceremonies of the hand-shaking sort are often assumed to be a recent addition to human life, probably because they are so very unlike anything an ape ever does. But human cooperation and the break from ape society is millions of years old, and *Homo* has always needed to find ways of holding communities together, maintaining trust, and imparting a serious sense of duty to the community. We do not need simply to visualize an agreement; we give it a physical grounding that can be recalled.

In this diorama, Dakila and his friend agree to stand guard while the other drinks from the pool. There is no historical evidence for such a ceremony, but the scene stands for whatever was done to create trust and duty. Although it is not as coherent as a true sentence, the ceremony does express what we can call a proto-idea. If an idea combines two or more points of attention into a whole, a proto-idea puts two points beside one another without their becoming a new whole. The same lack of a whole can be observed about protolanguage, which is a general term for any speech that precedes full sentences. All sentences must be organized according to rules, but protolanguage works more like a collage in which things are placed side by side and observers make what they can out of the relationship. In proto-ideas and protolanguage, points of attention are strung together without the expression that turns them into a unit.

> Dakila and another male see a young female. Dakila says, "*she*," and smilingly waves his two hands in and out in the shape of a woman's torso. The other male smiles.

Here is another way of turning two points of attention and their associations into a unit, again using a word and gesture together to express a proto-idea. The word *she* identifies a topic by directing attention toward a particular female. The gesture illustrates an answer to *what about the topic?*—she has an attractive shape. From a grammatical perspective, the important thing is that the two parts do not combine to form a unified idea. Despite the crudeness of the thought and the expression, Dakila's comment is more sophisticated than any phrase uttered by Gärd.

A gesture in a situation like this scene is full of ambiguities, and it is unlikely Dakila could answer verbally if asked, "She what?" Susan Goldin-Meadow has studied signing and gestures for years and has made an excellent little film of people using gesture to amplify and ground their point.

> A film clip on a monitor shows a congenitally blind woman giving directions. As she describes how to turn, she demonstrates the actions with her hands. "Turn left," says the blind woman and waves left with her hand.
>
> A second monitor shows a man making an abstract point about negotiation. He uses his hands artfully to clarify and concretize his abstractions, helping to establish a physical and spatial relationship for his words.

Goldin-Meadow stresses that the gesture expresses specific semantics, not its syntax. The difference between word and gesture should, by now, be clear to everyone. Gestures illustrate; words point. The pointing builds up an expectation. When Dakila says *she*, the listener naturally wonders, *What about she?* Gesture puts an end to expectation, but it can also satisfy. In this case it provides an answer, showing us something about *she*.

The ordinary *she* holds more promise than the eye-catching gesture because once people are capable of putting points of attention together, they expect pointing to be taken whole. Iskuhi could easily have said *she* and let the rest be understood, but more was expected by Dakila's time, even without sentences.

Some scholars, Terrence Deacon and Derek Bickerton among them, hold that words are a unique linguistic universal. It is not merely that a word like *she* is symbolic. Every language is likely to have some equivalent word, and in every case it produces similar expectations. Every language will require that you tell something about this *she*.

Gesture, however, carries no such expectation. A sign indicating a woman's curves may be widely, even universally, used, but no further word or gesture is required. When a statement combines word with gesture, it is the word side that can evolve to replace the gesture. Not the other way around. (Of course, we are talking about conversational gestures, not sign languages in which gestures are used as words and provoke the same expectations as words in spoken or written form.)

Capac | 1,550,000 Years Ago | Generation 361,550

Capac and another male are beneath an acacia tree, looking up at a long limb where a leopard has carried eight animals, mostly gazelles. The animals dangle upside down, pressed in joints between the big limb and its smaller offshoots. Capac says, *"up there,"* and his companion adds, *"much meat."*

The two phrases combined almost make a complete sentence. Still missing is a word that can tie the phrases together in a knot. One choice for a word might be *hangs* which makes a clear, visual statement. A less interesting choice would be *is*, but it does work. Many languages allow speakers to omit *is*, so Capac and his friend together have created a perfectly acceptable sentence for many speakers.

The fact that two people together can create one sentence forces us to consider where language, or culture more generally, is found. Does it sit out in the individual brain, or is it scattered piecemeal across the whole speaking population, or does it only exist when it is used? Many systems produce real phenomena that are not found in the system's elements. A woodland, for example, is an ecosystem with many local peculiarities. A food web in the woodland is the result of a great many interactions; it would be absurd to try to find the web in the individual members.

Anthills too are the product of interactions. The colony members do not carry the rules of the interactions in their tiny nervous systems, nor are the rules the sum of all their individual rules. Instead, the nature of the colony emerges from all the little interactions that take place. These kinds of undefined systems reflect the way they have no king. There is no central authority laying down the rules or controlling behavior. Language and culture strike many observers as behaving in the same way.

While the creation of one sentence by two speakers is unusual, it is not unknown, and many aspects of speech are as likely to arise from the setting and from interpersonal reactions as from the workings of the speaker's brain. One person may use a phrase simply because somebody else used it. There is a whole branch of study called sociolinguistics that examines language from this perspective. Other observers, with Chomsky in the forefront, argue that a language is simply the sum of what is in the brains of its users. They distinguish between the I-language, which is internal to the speaker, and the external E-language, which is the public summation of all those little I-languages going on in the brains of speakers and writers. Chomsky even insists that E-language is of no interest because it gains nothing from being public that was not already in the speakers' heads. The logic for this position comes from an analogy between human societies and computer networks. If human brains are a form of computer and human society is a type of network, why wouldn't brains, computers, societies, and networks all follow the same laws? In a computer network, information may be distributed throughout it, but each computer has its own internal rules for processing the information. If there are differences between internal rules in different computers, the result is an incompatibility between systems, not a growth in computer wisdom.

This analogy suggests that all languages will be fundamentally alike or that there will be mutually incompatible languages. Experience shows that unless they are very closely related, languages are mutually incomprehensible and require a translator. So there is some reason to doubt that all languages are fundamentally alike, but mutually incompatible languages would not even allow for translation and cannot work together. Mutually incompatible languages are a commonplace of computer science but are unknown among humans. If there were such things, it would mean that a newborn infant would not be able to learn any language in the world. There would be some languages a particular infant could not learn because the languages rest on a genetic disposition absent from those babies. However, colonial empires and global migration have shown that all babies can learn whatever language they need to learn, no matter what language their parents spoke.

A colorful map of the borough of Queens in New York City shows the 138 different languages spoken there. Different colors mark areas of heavy concentration for Chinese, Korean, Hindi, Italian, Greek, Russian, Tagalog, French, and Haitian French.

The United States has migrants from all over the world speaking hundreds of languages. The children of all of them turn out to be capable of learning American English. There are no inborn language incompatibilities. So, goes the argument, as there are no mutually incompatible languages, all languages must be fundamentally alike.

The argument against mutually incompatible language is so strong that linguists were stunned when a genetic study found that there are genetic differences between populations that speak tonal languages and populations that don't. Tonal languages use pitch to distinguish between words. All languages make distinctions based on differences in formant (energy concentrations), such as *toot* versus *tooth*. The *too-* in both words have the same formant, but *-t* and *-th* vary enough to be heard. Tonal languages have those kinds of distinctions, but they can also distinguish between *toot* and *toot* by saying one with a high pitch and the other with a medium or low pitch. Mandarin and Cantonese languages both distinguish many words this way. The Masai language of East Africa uses tone syntactically to distinguish, say, a sentence's subject from its object.

Linguists have always supposed that tone was just one more feature that some languages have and some do not, just as some have a future tense while others do not. But a remarkable study by two linguists, Daniel Dediu and Robert Ladd, rocked the linguistic world by reporting a connection between genes and tonality in speech. Even more startling, the distinction was between the genetics of whole populations, not individuals.

For example, Mandarin speakers as a whole (call it E-Mandarin) are likely to have a particular gene known to influence brain development, but individuals who think in Mandarin (I-Mandarin) may not have the gene. This finding contradicts not only Chomsky's expectations, but everyone's. It surprised linguists as much as if somebody had discovered that if most of your neighbors have brown eyes, then you have them too, even if you have only the genes for blue eyes. It suggests that for some language features, at least, culture can override the brain's own bias. We knew that culture could produce some such artificial effects. For example, there are many more blonds in American society than there are women with genes for blond hair, but nobody guessed that altering the way your brain works for language might be as easy as dying your hair.

Dediu and Ladd's pioneering work have complicated early *Homo* history. We can no longer assume that a change did not become population-wide until the supporting gene or genes became fixed in the whole population. Brain growth is ambiguous, but at least it leaves fossil evidence. A behavioral change without an accompanying genetic change is even more obscure. But it does suggest an explanation for the peculiar way the human species has such a mix of tastes and talents. For over a million years our lineage has been sheltered from the selective forces that spread genes throughout a population. Our brains are plastic enough and smart enough to get by without requiring genetic matching.

Capac and two others are on a koppie. Capac peers over a rock and says, *"them,"* and a moment later adds, *"go yonder."*

It is not fluent, but it is an elementary sentence. Two points of attention (them, yonder) are tied together by *go*. They have moved beyond ape capabilities. No signing ape has come close to signing even so simple a sentence.

Previously we have seen that speech builds on perceptions, directing attention. This added step of joining the points of attention creates a new perception, but this time it is active. By themselves, *them* and *yonder* point to static things, but the addition of *go* brings the perception alive, just as a movie projector, or the zoetrope we examined, makes a series of still images dance.

The value-added result of such liveliness is understanding. When a connection between sets of associations is truly unexpected, the act of understanding can produce a physical reaction, called variously "getting it," or "the penny dropping," or "the *aha!* effect." Thinking in a human manner—including the ability to put words into sentences—requires this capacity to link associations into chunks. Speech itself depends on it.

So, does navigating the world require joining associations? Take, for example, that sentence *The two leaning trees can be seen from the spring.* It is a quiet sentence, almost a still picture, except that it puts an observer into the setting as well. Navigation works in just that quiet way. Suppose Capac knows a place where two acacia trees bend together to form a recognizable shape. He might associate that sight with climbing a nearby low hill that screens a bushy area where berries can often be found. These associations are all tied to the sentence's first focal point, the leaning trees. Perhaps Capac also frequents a spring where water emerges from some rocks. He too knows the area around the spring, just as monkeys and rodents do. There is the second point of attention. But we need a knot to tie the two points together. Perhaps Capac is at the spring looking across the plain, and suddenly recognizes in the distance the distinctive shape of those two acacia trees associated with a low rise and berries behind. Recognizing the acacias and recalling the many associations connected with them, the savanna makes sense in a way that, just a moment earlier, it did not. Now Capac understands how to get from the spring to a promising berry patch; it seems fair to say he understands, because he has never made the journey that way before and has not learned by doing or by observing someone else do it.

Psychologists call this kind of sequential understanding "chunking"—the combining of two sets of associations into one chunk of information. It would be hard to believe that apes and monkeys, with all their three-dimensional travel through the forest canopy, do not have the ability to form navigational chunks. So why can't they form linguistic ones as well? The puzzle seems less enigmatic when you examine the "knots." For navigation, the knot that ties the places together are the two trees that can be seen both from a nearby ridge and from a spring, but the knot for making sentences is a verb: *can be seen*. In the diorama too the knot is a verb: *go*. Actions tie sentences together, while landmarks organize space.

Verbs appear late in children's language, at least most verbs do. Researchers into children's speech refer to a paradox. A few verbs, such as *eat* and *go*, form a regular part of any child's earliest vocabulary, but verbs as a class appear only in the third year or even later. The third year is the point where children's language outpaces ape signing, not just in fluency and spontaneity (which is apparent from the beginning) but in every capacity—socially, grammatically, and creatively. The sudden use of many verbs indicates another difference between apes and us—we understand the world as an active place. Apes don't tell us how they understand the world because they don't use verbs.

> Capac faces eight other adults on an open grassland. The horizon is cut short by an escarpment wall that runs along its whole length. Capac is pointing toward the horizon, but, except for the wall itself, nothing special is apparent. He tells the group, *"wildebeest go round,"* and waves his hand in a circle to report that he has found an animal in trouble.

In Capac's time, *Homo* still had many unpredictable changes ahead of it, but the outcome begins to look foretold. *Homo* is forcing its way up the food chain toward the rank of top predator. Already, *Homo* was not merely the smartest animal on the savanna; it is markedly smarter than it was only 150 thousand years before, in Gärd's time, when two-word phrases were the most advanced utterances possible. Yet Gärd would instantly recognize the ecosystem where Capac lives.

Brains by themselves do not determine one's rank as a successful predator. More important is strength, capacity for butchering, and, for the lesser predators, courage. Thanks to their size and claws, lions do not need much courage. It is the jackal, which has to steal meat from under the lion's nose, who needs shrewd pluck. The human lineage needed courage too, limited as it was by a gross physical weakness and a complete absence of a capacity for butchery. Descended from omnivorous food gatherers, the human lineage was maladapted to scavenging meat, lacking claws, fangs, or even the arm strength to rip chunks of meat off a dying or freshly dead carcass.

Homo increased its strength by hunting in groups, but group hunting is the norm among the big predators of African plains. Lions, hyenas, and wild dogs all hunt in groups. But *Homo* was getting smarter, and formed smarter groups. The others in Capac's group listen to his news of a turning wildebeest and see a detail of the African plain through his words. This group understanding puts *Homo* on a level of power unknown to other group hunters.

With the arrival of the sentence, it is as though the group had found a new, group-wise sense. What one speaker perceived, the others can perceive as well. A new, communal way of experiencing the world, a way known only to humans, has appeared. Sentences involve others in the understanding of co-speakers to an extent that is unimaginable to apes, or even to the dogs who live with us. The leap in understanding that comes by combining perceptions into action is so unlike what other animals do that they cannot grasp the difference. Animals can probably recognize *stronger-than-me* and step aside, or *faster-than-me* and keep their distance. But *smarter-than-me* goes beyond an animal's power of perception. In time, animals will respond to this mysterious difference with an undifferentiated fear. The results can sometimes take on a humorous quality as when an angry elephant backs away from a man it could kill in an instant, or a sleeping rhino suddenly jumps up and runs away, or a lion keeps its distance.

> Capac and a companion are facing one another on an open plain. A line of migratory herds fills the horizon. Behind Capac's friend, nearly hidden by the grass, is an immobile newborn Thomson's gazelle. Capac tells his friend, *"gazelle hide behind you."*

In trying to understand what animals know about other animals, philosophers and psychologists speak of a "theory of mind." Do animals have one? The mind is a concept of Western philosophy, so we cannot take the question literally. But many animals know that others can see things they do not see, and will follow the gaze of another if they notice it staring at something. There has been much effort to determine just what animals know about the "beliefs" of others. Apes seem to be good at knowing when they have seen something others in their group did not see. So the fact that Capac knows his friend cannot see a young gazelle is nothing new.

What is new is the way Capac has gone beyond recognizing another's knowledge of the world. Michael Tomasello points out that cooperative speech calls for a "bird's eye view" of relationships, which Capac demonstrates here. Capac and his companion each have their own perspective, but they also need to share a view that lets them see the two of them together. Capac sees both the gazelle and that his companion is looking the wrong way to spot the prey. An ape could do the same. Capac's sentence, however, does something unusual. Like much of the speech we have seen in this history, it points, but it also reflects the viewpoint of neither the speaker ("I see a gazelle") nor of the listener (who knows nothing of a gazelle). Instead it combines the two viewpoints and reports what the speaker knows (gazelle hide) in terms of the listener (behind you).

A feat like this may be the natural result of an ability to combine separate phrases, or it may require that ability and something further. One thing we do know is that it is a full stride beyond simply recognizing that others know other things. A bird's eye view seems of little use to apes and their individualistic society, but it must surely be helpful for members of a cooperative species to see themselves from outside, where they are part of a community.

> Two scenes are placed back to back. In the first, Capac approaches another man and begins vocalizing. The second man joins in. The two are not singing or even chanting, just making a kind of emotional bonding through vocalizing.
>
> In the second scene, Capac says to the same man, *"we go there,"* and points toward a koppie maybe half a mile away.

These scenes serve as a reminder that language still has a long way to go before it becomes a modern one. Simple sentences force speakers into abruptness. There is no syntactical ability to produce a more polite remark, something like, "A few of us are thinking of going over there. Would you care to join us?"

Nobody familiar with this more generous style of speech would be happy with a lifetime of Capac's bluntness, but in Capac's own day his bluntness would be taken for granted and produce no grudges. Still, the absence of any speech for furthering a sense of trust and appreciation must have been a problem for members of a community based on cooperation and mutual respect.

One solution would be to continue with the vocalization-based emotional bonding. Those bonds were already so old that Capac was as remote from their origins as he is from our own day. Indeed, even today, social encounters in our word-bound world commonly begin with culturally defined gestures, touching, and words like *hello* and *hi* that are barely more than friendly noises. Emotional vocalization of a more elaborate kind seems unlikely to have been replaced quickly by words.

Linguist James Hurford has noted something further about these basic topic-plus-news sentences. Their organization is "pragmatically motivated," that is, organized by the context of the speaker and listener's common ground, not by syntactical rules. Capac begins his announcement with *"we"* because the friend already knows who we are. If the friend was already focused on the koppie, Capac could say, *"there go we,"* once again moving from topic to news. It is this context-based choice of organization that enables Capac's group to speak simple sentences without having first developed some grammatical rules. Such sentences are convenient, but mercilessly abrupt.

Taranga | 1,549,930 Years Ago | Generation 361,555

Two women are squatting and watching their babies playing on the ground. One of the adults says, *"Taranga go."* The other replies, *"Taranga go where?"*

The linking of topics with news put an end to the acceptability of single-word or single-phrase declarations of a topic. A quarter million years earlier, Shoichi could name a strange group and it was a great stride toward better cooperation, but partial sentences will never do in the age of multiple phrases. *Them?* *What about them?* Speakers are expected to organize their utterances into intelligible combinations.

One of the seemingly simple facts about sentences that has drawn much comment is that some words are acceptable and some are not. "Taranga go blue house," for example, is crude, but we get it. "Taranga go blue lines," however, is simply confusing. It used to be considered self-evident that you could not say "president go blue lines," but since the invention of the computer, the unacceptability of many sentences has become an important point. Along with a definition, each word entered into a machine dictionary must be given a set of properties. Words have semantic properties (I can *stay put*, but I cannot *go put*), categorical properties (*go* is used as a verb but not as an adjective), and grammatical properties (*go* can be conjugated but not declined).

If you take literally the idea that your brain is a computer, it is quite hard to imagine how we learn all the properties of words, and even harder to propose a way to evolve a machine that organizes all these properties. Fortunately, we do not have to believe that our brains work just like the computers that are sold today. Contemporary computers do not pay attention and do not perceive—that is to say, they do not do the things that are central to the production of speech.

There is ultimately a common ground shared by all humanity. We call it the real world, and it constitutes the givens of human experience. Within broad limits, we all perceive the same physical reality of actors, actions, and objects of actions, played out in space and time. Those perceptions shape the way we combine words, and not the other way around. When language directs our attention to something, as in "Taranga go," we know whether we have been given a whole idea or not.

How do we know? Fortunately for this tour, we can take perceptual knowledge for granted as it rests on powers much older than the split between chimpanzee and human lineages. Anybody who has ever owned a dog has seen how it becomes confused when something disappears, and how it looks around to find it again. Perception and its assumptions are very old.

An enduring puzzle for late night discussions in the college dorm is whether the world we perceive is really out there, but the answer is apparent for the evolutionary-minded. Natural selection would have wiped out the primate lineage long ago if our perceptions did not provide a reliable guide to the moment that envelops us. We did not have to evolve the complex set of categories used by computer scientists to approximate the world around us. We just had to evolve the ability to keep directing our shared attention, so that listeners could perceive what speakers perceive. When a listener says "Go where?", the speaker is reminded to finish the utterance. This on-the-fly editorial process persists to this day and surely has been with us at least since the beginning of sentences.

Taranga plays with her baby, holding the infant up and saying, "*mama go far. mama find figs.*"

Ljiljana Progovac lists the action of placing sentences together without using any connecting words—known to grammarians as parataxis—as the simplest way to state a complex idea. Each part is a simple sentence. More complex versions of the same idea are *mama go far and she find figs* or even *mama*

find figs when she go far. Taranga is still speaking a very sim-
plified language.

> Taranga walks beside a half-grown child while a group of three gazelles
> bounds away from them. The child points and says, *"gazelle go away."*

The child in this scene is producing a good simple sentence, but why? There is no news here. Everybody can see what
is happening. But, as Jean-Louis Dessalles stresses, "Language
activity is a genuine urge." As they grow older, people take
some control over this urge, but if they are subjected to a long
period of forced denial of speech, it will erupt in a manner
typical of biological frustrations. "A natural behavior whose
expression has been prevented will at length be executed, even
though it may be irrelevant."

We saw that even before words, there was an urge to vocalize. Probably once language began, the urge was modified to
become an urge to speak words. The same thing has been observed in birds. They respond to the sounds of their "tutors,"
older birds whom they can hear, and try to master their songs.
Presumably, as with almost everything else in humanity, there
is quite a range in this urge's power. Some people are compul-
sively garrulous while others are content to go without speaking for long periods. But we must not forget that even in these
scenes where cultural learning is evident, biology continues to
play a strong role, especially in pushing children to master the
speech of their environment. That instinct to use the language
they hear, combined with the continuing ratification process
of using words in a way that comes easily, ensures that the
language continues to be mastered without much effort by the
youngest generation.

Olatunde | 900,000 Years Ago | Generation 419,250

Another wet/dry phase began and persisted. It was the first such period in almost three-quarters of a million years and it would turn out to be the last of them. Nine hundred thousand years ago, wet places were still scattered like strange oases amid a generally harsh, dry land. Large deep lakes appeared in what is now southwestern Ethiopia, central Kenya, and northern Tanzania. This was also the time of the mid-Pleistocene revolution, when ice ages suddenly became more frequent and the period between them shortened considerably.

Previous wet/dry phases introduced new species to the human lineage and its kin; however, it is a testament to the adaptability of *Homo erectus* that the final wet/dry phase produced no new species of *Homo*, although a somewhat mysterious branch of bipeds, the genus *Paranthropus*, did go extinct at this time. Presumably the stability on the part of *Homo erectus* reflected the same processes we see today—cultural changes that permit individuals to adapt without changing their basic biology. Cultural evolution moves so much more quickly than the biological form that genetic changes lag well behind. The survival of the *erectus* species throughout the long two hundred thousand years of a wet/dry phase is strong evidence that, despite the steadiness of the Acheulean tool kit, there were serious cultural changes underway.

This history is concerned with only one of those cultural changes, language. Language evolution accompanied cultural development from the beginning, but verbal culture could not become particularly interesting until speakers got beyond the single-phrase level. Even today the content and spirit of speech by toddlers around the world is quite similar, no matter what their language. But once speakers begin putting phrases together into topics and news, the great diversity of languages that we know today became inevitable. There seems to be an infinite number of ways to describe the world we perceive, but a limit on the number of ways people can think about their perceptions. Thus every culture describes the world a bit differently, choosing only a part of the very rich number of solutions possible.

We are not yet done with important brain changes in our history, but it seems almost certain that once speakers began combining phrases into sentences, language evolved quickest through changes in the culture. When vocalizing began, it was entirely biological, the result of a natural urge to make sounds. By the time of the first word, some culture had entered the story. In making a word, there was no biological reason to prefer one sound or combination of sounds from a range of possible sounds. Even so, it was biology that kept the species vocalizing and making words. Phrases made by combining words marked a cultural change, but it probably required some serious rebuilding of the brain to start combining phrases into sentences. This co-evolution of brain and language persisted, and perhaps continues to our present day, but the balance shifted from the brain doing most of the evolving to culture doing most of the heavy lifting.

For our purposes, this new dominance of culture makes the history more obscure and the dating of events even harder. Language does not leave fossils, but other things do, and a limited chronology is possible so long as there is a correlation between behavior and biology. But that chronology is trickier when there is no biology to guide us. Late *Homo erectus* cultures seem unlikely to have had contrasts as sharp as those found in the Amazonian basin today, but there may have been considerable differences between lives lived and the languages spoken around the lakes of Tanzania and those in Ethiopia.

This chapter is dated at nine hundred thousand years ago, chiefly because after two hundred thousand years another wet/ dry phase was coming to an end. The survivors of that period were the descendants of people who several times over the forgotten centuries had adapted successfully to sudden sharp changes. It seems likely, therefore, that their culture had become much more subtle in its capacities. Previous dates in this history were based on, at least, good circumstantial evidence, but in this case we cannot yet say much more than it looks like a good bet.

> Olatunde and two other males are in a woodland on the slopes of a steep cliff wall. They are seated, picking figs and chatting as they do. The scene is reminiscent of the one from over five million years earlier when Sara sat out on a limb plucking figs, except Olatunde is talking and one of the males is offering another a fig.

There is a paradox of sorts in the development of the talking species. On the one hand, language requires an ability to stay focused on some external topic. At the same time, speech demands the mutual engagement of speakers. There is a kind of double attention required, as when people talk together while doing something else. In this scene, for example, a group has some simple chatter going, maybe about the figs they are picking—*This fig fat juicy*. That sort of thing. They are also paying attention to one another's presence. We all know from our own experiences that this divided attention can overwhelm someone, so that awareness of the listener breaks the speaker's concentration and the speech is lost, or sometimes the speaker forgets the listener and concentrates so much on the topic that the speaker says things that are insulting or inappropriate for the listener, or sometimes the listener's attention wanders away from the topic and the speaker's words are missed, or sometimes the listener's attention focuses so much on the topic that awareness of the speaker is lost and the listener may respond inappropriately with, say, unsympathetic laughter. This tension between attention to topics and attention to others is very old, an inevitable social response to the speech triangle, once the ability to express complete sentences about topics became routine. In many ways, the members of Olatunde's generation were remarkably like us today, even though they lived about fifty thousand generations ago.

We still have not come to a point where people evolve special modules for handling syntax, despite the use of full sentences. Pragmatic habits, taking context and common ground into account, still go a long way. There is also what is known as the lexical turn, the associations and expectations evoked when particular words are used. At every step of the way, language cannot get ahead of what children can learn.

Just for fun, imagine that Generation N speaks a rich language that is too much for the next generation to master. Along comes Generation N+1, learning what it can, changing what it cannot remember, and improvising as necessary. The result is that Generation N+1 speaks a language that is much easier to learn. This logic has been demonstrated by Simon Kirby, a language evolution scholar in Edinburgh, Scotland. He generated a user-unfriendly language and taught it to student A, who then taught the language to Student B. Next B taught it to

Student C, and so on to the tenth generation, Student J. By the time Student J had to learn it, the language had become much easier to learn and much more regular in its sounds, and—apparently out of thin air—it had acquired a regular syntax.

The details of this experiment can be criticized because students A through J all had modern brains, so we cannot know how generations A, B, and on to J following Olatunde's generation would have learned. But the principle holds fast. Each generation learns what it can, alters what it can, and forgets the rest.

Thus, the Dick mystery from earlier is not much of a puzzle. How do children learn language easily? The process is very similar to evolution by natural selection. What survives to the next generation is whatever is best adapted to the environment. After a few hundred million years of such selection, the organisms are so well adapted that they seem designed for the place, or perhaps the place was designed for them.

Language selection works the same way. The words and phrases that are most used are those that are best adapted to the needs and capacities of speakers and listeners. After two million years of adapting, language seems designed for us, or were we designed for it?

In both cases, the design is an illusion brought on by repeated selection of variations.

The solutions to these two mysteries—Dick and Harry—rest on the same principles. Other species do not combine enough trust, cooperation, and intelligence to speak; we start talking, however, because we are so dependent on intelligence, trust, and cooperation that language just naturally blossoms among us like toadstools on a wet log.

Olatunde stands guard while four companions lie on the ground beside a small stream and catch drinking water in their cupped hands. Olatunde cries out, *"two lion jump kill zebra."*

Olatunde's cry is much more sophisticated than Taranga's *"mama go far. mama find figs."* In those remarks, two sentences brought news about a single topic (mama). Olatunde's report also brings two bits of news (jump, take zebra) about one topic (two lion), but Olatunde did not say *two lion jump zebra; two lion kill zebra.* Nor did he use a conjunction to say, in the modern way, "Two lions jumped *and* took a zebra." The form used here comes from a type of speech seen in early Creole languages, chosen because they seem to offer the best examples of the syntax of very young languages. In those new languages, strings of verbs, known technically as serial verbs, of the *jump kill* variety are standard. Serial verbs are formed by stringing together two distinct verbs that share a common subject. Serial verbs of this sort are rare in mature languages, but they appear in young Creoles around the world.

Like other verbs, serial verbs tie together two points of attention. In the example, they serve as a collective knot for *two lions* and *a zebra.* This shared function is possible because verbs do not redirect attention. The use of two verbs is possible because both share a common subject. The same steady focus permits English speakers to omit a second reference to the subject and say something like, *Two lions jumped and killed a zebra.* When attention has been redirected to another topic, the resulting sentence must restate the subject or become confusing. For example, try reading this clumsy sentence without including the second reference to the two lions: *Two lions jumped and a nearby gazelle ran away and the two lions killed a zebra.*

Serial verbs are striking for their vividness and many storytellers might do well to keep the technique in mind; some writers, in fact, do use them. Henry James could be especially flamboyant at stringing verbs together, as when he wrote, " . . . after which she had embarked, sailed, landed, explored and, above all, made good her presence." The word that gives away the language's modernity is *and*. Such a simple word is slow to appear in young languages. More mature languages also have verbs that compound actions. A serial verb might be *He take carry* . . . while a modern speaker combines the two into one verb: *He brings* . . .

The introduction of serial verbs shows how far speakers have come:

- Single words focus attention on one spot (e.g., *purse*).
- Phrases maintain the focus but broaden the perspective by adding a set of associations (*Sylvia's blue purse*).
- Sentences express an idea by using a verb to combine a topic phrase with a news phrase (*John take Sylvia's blue purse*).
- Serial verbs transform an idea into an episode (*John take carry Sylvia's blue purse to Mary*).

A prominent New Zealand psychologist, Michael Corballis, argues that the truly distinctive form of human thought is not symbolic but episodic thinking. An episode joins a set of facts or associations in time. Corballis distinguishes episodic thought from a generic memory in which a single memory is associated with an emotion, perception, or action. Generic memories are sufficient to permit simple stimulus-response conditioning. A sensation of some kind (perhaps a bell) triggers a generic memory (perhaps the image of food) that triggers a response (perhaps drooling).

Episodic memories organize many more associations and produce responses that are much less predictable. Episodes allow for thought about both the past and the future. Since we have already seen the past, it might seem that imagining the future requires special abilities not needed for remembering, but the two activities are closely related and depend on an ability to organize associations into a whole. Brain-imaging studies have shown that remembering past events and imagining future ones activate the same core network of neurons. People with amnesia as a result of brain damage have as much difficulty imagining new experiences as they do remembering past ones.

The essential problem for speakers is to organize an episode into a series of related associations by turning them into one or more full sentences. The role of syntax in this task is to ground a sentence's meaningful words (i.e., the words that pilot attention to a doer, a deed, or a done-to). Many linguists, led by the great Chomsky himself, maintain that syntax is a series of arbitrary rules for organizing arbitrary symbols, but that idea hit a crisis when people began to demand an explanation for how such a system could evolve. The linguists had no good answer. Instead, answers have come from other sources, like Terrence Deacon who put it very nicely: Syntax does the pointing when there is nothing to point to. Call it virtual pointing.

Languages scattered around the world have so many different ways of handling virtual pointing, and their methods often conflict. Other languages insist on pointing at or ignoring so many different things, that most, quite probably all, syntax is learned rather than inborn. And yet there is a profound similarity running through all this difference. The relationships and associations expressed by virtual pointing have to do with space, time, background, and foreground. A German linguist once observed that "Language translates all non-visual relationships into spatial relationships. All languages do this without exception, not just one or a group. This is one of the invariable characteristics of human language." These are the givens of perception, and the capacity to perceive is inborn. Other possible relations—such as logical, evolutionary, economic, Newtonian, familial, mathematical, or astrological—might be used, but they never are. Syntax does the work of perception. It works that way because it depends on joint attention, and attention is a power of perception.

The details of the evolution of syntax are lost in very deep time, likely irretrievably lost. But one day, possibly as long ago as Olatunde's time, members of the genus *Homo* began speaking sentences like "We caught a gazelle that ran blind from two lions." The topic is *we*, the news is *caught a gazelle*, but now there is context as well. The gazelle was fleeing lions. A sentence like this one contains the equivalent of a foreground (topic and news) and a background (context), and as we all know, background and foreground is a matter of focused attention. We can watch a person walk through a snowy field, or switch focus to the snowy field in which, as a background detail, someone is walking. Foreground/background sentences can always be switched so that the context becomes the topic and news, and vice versa. Thus, that sentence could be stated as "The gazelle we caught ran blind from two lions." With speech, this is flexible—it is up to the speaker to decide where to put the emphasis.

Why does this history propose that the date for such elegant speech might be as old as nine hundred thousand years ago? Chiefly because there was still much to accomplish before modern speech was possible, and not all of it was just a matter of one more cultural breakthrough.

GRAND HALL: BECOMING HUMAN

T IS TIME for a little wondering about this history itself. Has it gone down a garden path?

Given the history of language seen so far, how is it possible for people today to talk about abstract concepts? We do not perceive justice, yet we have a word for it and history has been made by people who demanded it. We do not perceive metaphors, but we speak them. Many cognitive psychologists and linguists, therefore, deny that there is a relationship between understanding speech and perception. As Derek Bickerton asks, if you think visually how can you understand a sentence like *"My trust in you has been shattered forever by your unfaithfulness"*? Here we have two abstractions—trust and unfaithfulness—linked by a metaphor, shattered. True, the sentence seems to work like a normal perception that combines two points of attention through a verb, but the points of attention are not perceptible and the unifying verb is a metaphor. The verb, you may notice, is in its passive form, making the whole sentence just a little more abstract and hard to visualize. Yet an easily visualizable sentence could have the identical structure: *My house in Alabama has been shattered to pieces by your bomb.*

So we have two things to wonder about. How is it possible to use concepts and metaphors? Why do we organize abstract and metaphorical sentences as though they were about perceptions?

This tour's answer is direct enough. Scholars must think in evolutionary terms. We began speaking about points of attention and unified perceptions. That was the basis of our language for probably a million years. Later, after the last of the wet/dry phases, after *Homo erectus*, we began speaking about things that could not be perceived. By that time, however, we had already developed a bio-cultural way of organizing sentences that depended on perception. If we want to speak about abstractions, we must do so as though they were part of the concrete world. Even if we use a much less vigorous sentence than Bickerton's lively prose—something like *Because of your unfaithfulness, I can no longer trust you*—we still see metaphors of concreteness. People do not possess unfaithfulness in the say way they do a jacket, but grammatically they are the same: *your jacket, your unfaithfulness*. Likewise, being able to run is physical while being able to trust is something else, but we say *can run* and *can trust* as though trusting and running were both motor activities.

Still, many experts suppose that from the beginning language has been used to express concepts, and that concrete expressions are just one subset of the many kinds of sentences people speak. Two prominent leaders in this approach are Tecumseh Fitch and Steven Pinker. Fitch is the author of the leading textbook on language evolution and Pinker wrote a bestseller called *The Language Instinct*. Both men are distinguished scholars who have immersed themselves in the data of languages and cognitive science, but this history builds on the ideas of an insurgent group that disagrees with the establishment.

The dissenters are scattered around the globe. An important group in Italy has worked out many of the semantic details of how language directs attention. The French linguist Jean-Louis Dessalles has made an important study of the relation between language and perception. British and American dissenters are plentiful too, although they tend to come at the matter from the flank, letting the traditionalists concentrate on formal sentence structure while the dissenters focus on pragmatics, the study of context's role in organizing sentences and how knowledge and interests shapes language. The insurgents' motto was summed up by James Hurford—a professor in Scotland who has persuaded many young linguists to study language origins. He began a presentation at a conference in Poland by asserting that "most syntactically interesting phenomena are pragmatically motivated." And a bit later in the presentation he asked, "If languages were not used by people to make statements, ask questions, and give orders, all with different assumptions about what the hearer knows, and what the speaker wants to emphasize, *what would syntacticians have to theorize about?*"

A deep controversy endures. On one side is an established curia that focuses on fixed rules and says language began by expressing concepts. On the other side is an energetic but less unified group of dissenters who concentrate on context, attention, and perception. The dissenters hold that metaphors, concepts, and mystics came after the expression of perceptions. This account unabashedly sides with the dissenters.

Most importantly, the pragmatics' path makes sense of the evolutionary data, and of the visible behavior of children as they start talking, and of the mysteries of language. We can take a random fact about children's language: The head of Rutgers' Infancy Studies Laboratory, April Benasich, reports that 5 to 10 percent of children have "specific language impairment," a problem with learning to talk despite having normal hearing, normal social abilities, and normal intelligence. Benasich thinks the problem is perceptual and notes that there is a strong correlation between perceptual discrimination in infancy and language ability at three years. By itself, it is one more lab result out of millions, but it does tie in exactly with what our history would expect.

Neurological data also supports the dissenters. Animal communications are triggered in the brain's emotional centers. When a vervet monkey sees a leopard, it has an emotional reaction that triggers a particular sound. Other vervet monkeys are alerted by the cry and very likely begin making the same emotion-based calls. A human seeing the danger is likely to be alarmed too and whisper "leopard," but the part of its brain supporting the sound is the sensorimotor region that handles perceptions and responses to perceptions. A neighbor, hearing the warning word, may be as terrified as a monkey, but the neighbor usually does not begin echoing the word "leopard."

Perhaps even more striking is the fact that we have created a number of artificial languages for working with concepts. Mathematics is thousands of years old, and more recently we have created a variety of computer languages that are used specifically for processing symbols. Natural languages and these artificial conceptual languages differ radically, so much so that the psychologist Gary Marcus has suggested that we did not do a very good job at evolving language and that computer languages show how much better they might be. He may be correct, but it could also be that computer languages and natural languages perform different functions.

All natural languages have three features, universal among perceptions, but not found in either mathematics or computer languages. First, languages focus attention on at least one thing while artificial languages have no such preferred symbol. Second, languages express a point of view; conceptual languages are proudly objective. Third, languages can change perspective, providing a narrower or broader view.

> A series of photos taken at a circus illustrate the features of natural language that make it so different from conceptual language. The top picture in the display shows a tightrope walker balanced on a wire and standing on his head. A spotlight makes him very bright, although in the dimness behind we can make out his assistant standing on a platform and looking on with concern.

Language is about something, but more than that, it focuses on something. You might say that $e=mc^2$ is about the relationship between matter and energy, but it does not focus on either side of the equation. Language is different. Take a simple sentence like *Those apple pies are mighty tasty*. The spotlight in this sentence is on *pies*. Why is *pies* pluralized instead of *apple*? Linguists argue over the answer to that one, but we can be simpleminded and say apple modifies pie and not the other way around. It is also possible to omit the word *apple* without changing the sentence's topic—*Those pies are mighty tasty*—but if you omit *pies* and keep *apple* the sentence gets a new topic. Notice too that several other words depend on *pies*. The sentence says *those* and not *that*, *are* and not *is*, because *pies* is pluralized. They agree with the spotlighted word, which is also inflected. If we inflected the other noun and said *Those apples pie are mighty tasty*, the listener would be puzzled. In the picture of the spotlighted wire-walker, other things are visible but the focus is on the headstand. In language too, dominant elements hold the focus.

> A pair of photos illustrate point of view. One picture shows the wire-walker from below. He is very high up, on a wire so thin as to be invisible, and he holds a long pole for balance. Beside that picture is a second, taken from above. The performer is much closer in this view and the wire is sharply evident. You can also see the special shoes the walker wears and, below, filling the picture's background, is a net waiting to catch the man if he falls.

All languages express a point of view, even when they are about concepts and non-perceptible events. One person might say, "She *sold her soul* when she lied about her beliefs." Another might say, "She *played her cards right* when she lied about her beliefs." A third, "She *dodged my booby-trap* when she lied about her beliefs." Each of these statements use an action metaphor—*sold*, *played*, *dodged*—to voice an invisible judgment about what happened, and each metaphor reflects the speaker's viewpoint. Even irresistible facts reflect a viewpoint. *The sun rises in the east* is true on earth, but an observer on the moon will note that *the earth rotates from the west* and someone on the sun might say *the earth is always out there*. Meanwhile, the mathematical laws of physics are the same across the whole of the universe.

A final pair of photographs illustrate perspective. The first picture
shows a close-up of the wire-walker's face. The concentration on it is
shocking; he is completely focused on his actions. The second shows
the same scene but with a wide angle. The platform where the walker
stepped off, the wire, and an assistant standing on the platform are all
visible behind the walker.

Language can zoom in or out in exactly this way. For example: *The president is speaking; The president is speaking
to a cheering crowd; The president, with the vice president
right behind him, is speaking to a cheering crowd.* We can do
the same thing metaphorically: *My hopes have run away; My
hopes and expectations have run away; My hopes and expectations have run away into the arms of my enemy.* Like a movie
director, a speaker chooses how broad a perspective to provide
the audience. There is nothing similar in computer languages,
animal signals, or mathematics. We cannot move out from
$e=mc^2$ to notice other details, nor can we zoom in on just the
m to learn more about it.

So why does anybody argue that language is about expressing concepts?

The traditionalists ask a basic question: Where do our topics come from? As the French psychologist Anne Reboul puts
it, speaker-listener interaction "presupposes content of some
sort, but . . . cannot account for content."

A naïve reply is that we look and we see, but that answer
certainly will not work for a computer. Consider a security
system that scans the faces of people in an airport terminal.
The faces are compared with those in a database and identify
known terrorists. It is very impressive, but depends on having

faces already stored in the database. If there's no content to begin with, there are no results in the end. As the well-known proverb has it, "Garbage in, garbage out." So if an empty database will not work for a computer, how can it work for our brains?

Some thinkers, notably the philosopher Jerry Fodor, have taken this argument to the extreme and said that everything we recognize must first be built into the brain, including very modern things like smartphones. The obvious retort to that position is that we cannot possibly have evolved brains that already had the concept of smartphones wired into them. Fodor agrees and says it is too bad for Darwinism. In fact, he rejects Darwin's theory of evolution by natural selection. But most concept-first supporters are less doctrinaire, allowing for a great deal of learning that builds on a few basic concepts. Steven Pinker, for example, speaks of a "conceptual semantics" in which "word meanings are represented in mind as assemblies of basic concepts in a language of thought." The words we use are determined by the concepts that are built into the brain; the rules for assembling concepts set the limits and possibilities of speech. The flow is from concepts to thoughts to language. Concepts shape language and not the other way around.

It is a perfectly logical syllogism: All computers begin with built-in concepts; all brains are a form of computer; therefore, all brains begin with built-in concepts.

What's wrong with that argument? One can suspect that there is something wrong with the computer-brain analogy. For example, computers do not need to pay attention to the input they process, while primates do. We pay attention to things that we do not know. We hear a noise. If we recognize it, we may ignore it. If we don't recognize it, we turn our attention to the source to identify it.

We may have no concept of a smartphone, but a friend shows us one. We pay attention and learn a bit about it. It seems appealing, so we buy one. By using it and paying attention to what we are doing, we learn more. If we cannot figure out something about our phone, we ask a friend who then uses words to explain to us what to do. We pay attention and catch on. Suddenly we know all about smartphones, when a year earlier we did not even know what they were.

This capacity to pay attention and learn distinguishes humans from the computers in use today. But let's face it, any attention-based explanation for learning depends on a total mystery. We have no idea how attention works, or how it is even possible. The psychologist William James commented on this ignorance more than a century ago: "Whoever studies consciousness, from any point of view whatever, is ultimately brought up against the mystery of *interest* and *selective attention* . . . [Consciousness's role is] always to choose out of the manifold experience presented to it at a given time some one for particular accentuation, and to ignore the rest."

The only defense one can offer is that science works by rolling back the mystery, not getting rid of it. There is always more to wonder about. We wondered how language could have begun. Then how words could have expanded into sentences, and now how concrete speakers developed the ability to use metaphors and abstractions. At the end of this quest, however, we will not find perfect enlightenment. We can still wonder about the particularities of perception. How is attention possible? Why are some sensations pleasurable while others are repellant? Why indeed does the stimulation of sensory nerves produce any sensations at all?

The consolation is that we are rolling back the mystery and replacing confusion with partial understanding. The idea that we first evolved an attention-based ability to vocalize—to utter words, phrases, and sentences about concrete things—resolves a number of mysteries that still baffle those who argue that conception came first.

One mystery for the traditionalists is how to fit the process into the theory of evolution. Not everybody agrees with Fodor that Darwinism has nothing to tell us about language origins, but traditionalists do tend to argue that language came late (perhaps only a hundred thousand years ago) and that it came suddenly. The many little adjustments of the vocal tract, brain, and lungs required to support language must have evolved for some other reason and then been put to use by speakers in a new way. It is not impossible—after all, we evolved our capacity to run and manipulate objects with our feet before we put those skills to use playing soccer—but nobody has any ideas about what use all those linguistic pre-adaptations served.

Noam Chomsky explains the need for a sudden leap in evolution to produce speech by insisting that without a modern language, one able to use complex, abstract sentences, a person would not be able to think deeply enough to gain any survival advantage over others with more ancient thought processes (that is, over perception). Thus, traditionalists favor the idea that language evolved in a very unusual evolutionary leap. We have seen that dissenters say language evolved in the ordinary, piecemeal way.

Another great mystery of language unsolved by the concept-first approach is the nature of meaning. A figure as important as Tecumseh Fitch has expressed great frustration that the decades of generative linguistics and the study of language origins have brought us no closer to resolving the enigma. He laments that "we have a good theory of information [machine communications] but we lack anything even approaching a good theory of meaning."

How does language get a listener to understand what the speaker perceives or thinks?

For the perception-first school, this question is not too mysterious. The speaker and listener pay joint attention to the speaker's topic. Whether the topic is present as a perception or as something imaginary, the speaker's words direct the listener's attention. Nouns draw attention to objects, verbs to actions, prepositions to spatial or temporal relations. But there is a problem with this explanation. It plainly does not work for today's computers because they cannot direct their attention. So if you insist that brains are a form of computer (as computers are currently understood), then meaning and attention cannot go together.

Divorcing attention and meaning, however, makes paradoxes pop up like springtime dandelions. First, people assume that if meaning is not the abstract result of joint attention it must be a real thing. It is common to read that "Language pairs sounds with meanings," which makes meaning seem as physical as sound. Then they start looking for that thing. Presumably that thing is somewhere in the brain, an assembly of neurons of some sort, though just where the assembly sits remains mysterious.

Equally mysterious is the way meaning seems free of any particular language. The English *Here comes John*, the Swahili *John anakuja*, and the French *Jean arrive* all pilot attention to the same phenomenon. But if you are caught up in the idea that meaning is a physical thing, you will assume that the brains of the speakers of all three languages must hold the same meaning within their neurons. The same meaning underlies these three sentences. The brain then becomes a computational system that translates its internal meaning into a string of publicly available symbols. This process of making internal meaning public is known as externalization. The symbols chosen depend on the language known to the speaker. There is, of course, no agreement over how these translations occur.

Another puzzle emerges immediately. If the meanings are universal, why do the symbols vary so much? And here is one more puzzle: If we have universal meanings in our brain, why does anybody ever think in words? We translate our meanings into symbols and then retranslate them back into meanings as though we were packing boxes, moving them from one side of the brain to the other, and then unpacking them again.

Most ironic of all is that this tangle of spooky mysteries has been created in the name of objective realism. The relation between attention and learning has been known for thousands of years, but attention strikes some scientists as too subjective a phenomenon to be allowed explanatory powers. So they end up relying on some ghost meaning that is somehow paired with the sounds of language. There is no way to detect that ghost, but like earlier scientific dreams such as phlogiston and the epicycles of planetary motions, logic seems to demand that the physical meaning must be there.

So this account is standing with its perception-first history, but that approach still leaves us uncertain about the dates for the appearance of metaphors and concepts. Dates are always difficult, but these are so uncertain that we cannot propose even general ones. Thus our tour can no longer present individuals as representatives of the generations from around the time when specific changes came into place.

> Three *Homo* with flatter faces than the ones in the previous hall are squatting around a fire. One is throwing some dry dung into the flames. A fourth *Homo* is approaching the fire, carrying more dry dung.

We do know that by eight hundred thousand years ago (about 425 thousand generations after Sara) the human lineage was showing signs of progress. The earliest physical trace of controlled fire comes from about this time at an Israeli site called Gesher Benot Ya'aqov. The first evidence of the division of labor comes from the same period.

Fire's discovery is the stuff of myth, a secret stolen from the gods. The Prometheus story of how a hero passed on the secret to humans is told in so many cultures that it seems a good candidate for a story known for tens of thousands of years. How clever did our ancestors have to be to put fire to use? The trick must be harder than it looks if *Homo* existed for two million years before seizing upon it. Of course, there have been fires ever since free oxygen became plentiful on earth, and it is quite possible that a fire that looks natural was set on purpose. Controlled fire may be much older than eight hundred thousand years.

What kind of thinking is required to set a fire? It seems as practical as toolmaking. Perhaps the difference is in the kind of thinking fire inspires. Making a tool can inspire interest in craftsmanship, while people sitting around fire can think about anything.

Another prehistoric find, this one also from the Middle East, at a site near the Dead Sea, dates from that same period and seems to show a division of labor. A *Homo* settlement— got that? An eight-hundred-thousand-year-old settlement!— shows that flint was worked in one location, shellfish were cracked open in another, and vegetable remains were left in yet another spot. It is tempting to say that if they could think so exactly about organizing a settlement, they must have been conceptualizing. But there is no evidence of a chief or dominant leader who sorted the work out this way. There could easily have been a community where people worked together according to their own personal inclinations—no chiefs or concepts necessary.

A third great find from eight hundred thousand years back is a fossil bone, the hyoid, that attaches the muscles which manipulate the mouth and vocal tract. This little bone is the only part of the vocal tract with a hope for fossilization. The rest— the tongue, larynx, vocal cords, and so on—are all too soft to survive. The hyoid, however, is a real bone and occasionally is found as part of fossils. The hyoid's human version differs from the generic ape version. In humans, the hyoid is shaped like a horseshoe; in apes, the bone has a cup-shaped feature in its middle. Apes anchor muscles from their laryngeal air sacs to this cup. The change in the fossilized bone's shape is evidence that by eight hundred thousand years ago the air sacs had disappeared from the human lineage.

The Dutch linguist Bart de Boer modeled the effects of air sacs on vocalization and found that they result in lower speech tones and a smaller range of possible vocalizations. De Boer hypothesizes that when content of vocalizations matters more than the impression made by the vocalizations, the air sacs would disappear. If we had enough hyoid bone fossils, we might be able to focus very precisely on when in the human lineage such a crossover occurred. Unfortunately, the next oldest hyoid fossil dates from about 3.3 million years ago, preceding the appearance of *Homo* and communal cooperation by half a million years.

By eight hundred thousand years ago, various *Homo* groups were scattered across Asia, Europe, Indonesia, and the British Isles. Some had fire, others not. The Acheulean axe was widespread, but it never reached East Asia. Labor appears to have been divided in the Middle East, but perhaps not everywhere. Distinct communities surely had different vocabularies and grammatical systems. In other words, cultural differences already shaped much daily life, and these were not just cultural differences in the philistine sense of being learned traditions rather than genetic behaviors. That level of culture is part of the ape world, as in cases where one chimpanzee group uses twig tools and another uses stone. Those uses support purely natural activities like obtaining food. A glance at *Homo* today shows something utterly different: the use of cultural artifacts like paint pots, necklaces, and prayer beads, whose functions can only be explained in terms of other elements of the culture. Chimpanzee culture is an adaptation to nature; human culture is an adaptation to culture itself. During the past eight hundred thousand years, this more profound form of culture has been growing, slowly at first but ever more quickly, like a honeybee picking up pollen. And for culture to grow, language must grow.

Linguists like to insist that every language is suited to its culture's needs; poets and other writers tend to disagree. Poets are the crafts folk who feel the limitation of their speech most bitterly because they feel the need to say things that cannot yet be said. It was no doubt this pressure that moved speech beyond talk of perception. Concepts eventually appeared, but they were likely very late. Between perceptions and conceptions stand metaphors.

A linguist, George Lakoff, and a philosopher, Mark Johnson, coauthored a classic work titled *Metaphors We Live By* that stressed the importance of metaphor to thought. Most scholars before Lakoff and Johnson considered metaphors a kind of rhetorical device for comparing two things while suggesting they were the same. In an age of analysis, rhetorical devices did not seem very important and few scholars paid much attention to their use. Statements like *His landlord is a fire-breathing dragon* or even Homer's famous repeated references to *rosy-fingered dawn* can have a humorous feel of blather. We all know dawn does not have fingers of any kind and landlords are not really dragons. But Lakoff and Johnson argued that metaphors are much more than literary flourishes. They are critical to human thought and cannot be done without. Metaphors "create a reality rather than simply . . . give way to the conceptualizing of a preexisting reality." In other words, if we were to move from the perception-dominant world of eight hundred thousand years ago to the conceptually dominant

ones of our present epoch, we needed metaphors to untie us
from the here and now.

Metaphors work by grounding abstract concepts in something concrete. Take Bickerton's sentence about trust being
shattered by unfaithfulness. *Shattered* is a metaphor. Trust is
not really like a clay pot that can be broken into little pieces.
We could use other metaphors: *totaled*, as though trust were
an automobile; *evaporated* makes trust a liquid; *nullified* turns
it into a legal relationship. But whatever verb we use must be
a metaphor. If we try to invent a non-metaphor, perhaps *koroxilate*, and then define the word, *koroxilate is the mental
process that _____ trust*, we circle back to the same problem.
Whatever word we use—*shatters*, *evaporates*, etc.—to fill in
the definition's blank will be a metaphor. If we were defining
something perceptible, we could replace the blank by pointing
to what we perceive, but there's the rub. In this case, there is
nothing to point at.

Metaphors take their grounding from resemblances. We
can speak of shattered trust because we accept that there is
some resemblance between a shattered object and what has
happened to our trust; however, not just any verb will serve.
We cannot speak of *gargled* trust, or *slapped* trust, or *carried*
trust, at least not in this context. The speaker has to spot a resemblance and the listener must also recognize it. Homer was
clever to spot a resemblance between dawn and rosy fingers,
but his audience had to grasp the resemblance as well. When
did a sensitivity to resemblances arise?

The study of written languages goes back only five or six
thousand years, and during the whole of that period, changes
in metaphors have been cultural. But there does seem to be a

biological side to metaphor. Children with autism commonly suffer from a severe inability to comprehend metaphors. One suggestion is that differences in linguistic processing may be at fault. Although much language use depends on the left side of the brain, metaphor comprehension appears to require a strong contribution from the right half.

This biological capacity to use metaphor likely evolved late in the human lineage, sometime during the last eight hundred thousand years. Its absence or severe deficit in autistic people suggests a function for metaphor. Autism disorders come in many forms, but they are united in their impediment of social understanding and interaction. Perception shows us only a person's exterior. Another's thoughts, feelings, and sensations are permanently inaccessible through the senses alone. Metaphors offer a solution. Even if you yourself have never been betrayed by a lover, the violence and destruction felt by another is revealed in the metaphor of shattered trust.

It is with the rise of metaphor that language accomplishes one of its greatest wonders, the way it enables speakers to reveal intimate details that otherwise would be forever unknowable. The Sanskrit scholars of ancient India had a great understanding of language and syntax, and in their tradition, intimacy was language's grandest gift. A Sanskrit hymn sings of language's birth:

When, O Lord of the World, the Wise established
Name-giving, the first principle of language,
That which was excellent in them, that which was pure,
Hidden deep within, through love was brought to light.

Here we see the basics. Language brings something "hidden deep within . . . to light."

With the appearance of metaphor, speech becomes something far more personal than it was during the long age of *Homo erectus* when speech served as a tool for sharing perceptions or concrete thoughts. Being human was starting to take on its modern sense of knowing and caring about others.

> Two young *Homo* look up toward a tree where a honey bird sings. If they follow the bird, they will find a hive filled with honeycomb. For its trouble in pointing out the sweet treasure, the two should leave the bird a chunk of comb to enjoy.

These two individuals represent the siblings of the last common ancestor of Neanderthals and us. Each of the *Homo* in this diorama will branch into a distinct lineage. A scene such as this one would benefit tremendously from a set of accurate chronologies. First, when might it have taken place? A large team of scholars investigated that question based on anatomy and genetics and concluded it was 660 thousand years ago, give or take 140 thousand years. That puts it anywhere from 800 thousand to 520 thousand years ago, giving us plenty of leeway.

A second mysterious chronology is the rise of a biologically supported ability to use metaphors and concepts. How far along had that process gotten when the lineage split? Since we know little about the dates of metaphors, we are in a weak position to say anything. We do know that by the time of the split the lineage was already using language, and was able to speak clearly about concrete things. How much they were able to look into what was hidden deep within their neighbors or

themselves is a mystery. And even after the split, trends continued on both branches. The Neanderthal's brain continued to
grow and even became slightly larger than the brain of *Homo
sapiens*. Neanderthal tools grew more elaborate, and eventually Neanderthals began using face paint and beads.

> A display table shows a series of stone axes and related tools. To
> the ordinary observer, they probably look like other Acheulean
> axes, but expert examination has found a break in the tradition that
> began making the first Acheulean axes a million years earlier. The
> displayed axes all reflect the "late Acheulean" tradition that began five
> hundred thousand years ago and required better motor skills, a better
> understanding of stone fracture, better planning, and more varied tools
> for making the axe.

By the start of the late Acheulean, *Homo* had been using
tools to make other tools for two million years. There must
have been some interaction between the stone toolmaking and
speech, but what? At a minimum, there must have been a specialized vocabulary for the striking motions of the hand, different kinds of stones, the different ways a stone can flake off, and
routine mishaps such as shattering a stone. These specialized
words are concrete. They point attention to perceptible things,
not abstractions, but they provide openings to metaphors. An
obvious one is the name for shattering a stone. The progres-
sion seems straightforward: *He shattered the axe* is a simple
statement of fact using a special word for what happened. *He
shattered his hand* transfers that word to another context, a
routine way word usages expand. Also routine is the blurring
of exactly what *shattered* draws attention to. A shattered hand
is different from a shattered stone. The pieces of hand are not

scattered about. The hand may heal but a stone never does. So the precision of where the word points becomes blurry. Later, someone might say, *She shattered his smile.* It seems like a giant stride toward metaphor, but perhaps it is simply a further blurring of *shattered*'s context and effect.

Given the way words change in languages around the world, a process of generalization of usage seems inevitable, even during the earliest years of speech. The *Homo* ancestor of a half million years ago (probably *Homo heidelbergensis*) surely had a more limited imagination than our own, and that could have slowed generalization. But by then the *Homo* brain was approaching its modern size. The more likely source of stability was culture's low center of gravity. The presence of few things to change led to few changes. Even so, generalization was bound to occur over evolutionary time.

A table displays the "Venus of Tan-Tan." It is a stone-dated to about four hundred thousand years ago (plus or minus a hundred thousand years) and stands about two and a half inches long. It appears to show a human form, two legs, a shortened trunk, and a large head. It was found in Morocco, near the town of Tan-Tan.

There is some debate as to whether this "Venus" is just a stone that eroded into an interesting shape or whether somebody long ago saw the shape, recognized its human-like qualities, and worked on the stone to highlight those recognizable features. It was found in a pile of stone tools, suggesting that it had been worked on. The most persuasive detail is in the "hands" found on both sides of the trunk. They are not generic stumps, but seem to include fingers. Such precise detail seems unlikely to appear accidentally on both sides of the figure.

Anthropologists tend to dismiss the object because, even if it was deliberately fashioned, it has no larger symbolic importance. As one anthropologist, Thomas Wynn, says, "If it is a one-off, I don't think it counts. It's not sending a message to anyone." But many artists are not trying to send messages. They are interested in shapes. Michelangelo used to say the shape was already in the stone; he just let it out. Maybe somebody three or four hundred thousand years ago could have said the same thing.

If we look at the Tan-Tan "Venus" from the perspective of where the lineage had come from, we can see the rise of interest in the speech triangle's third corner, topics. For over a million years, *Homo* had been using words to direct attention to topics. Other groups, gyrating wildebeest, ways to break stones—all of these things are practical topics, but the "Venus" is not practical, not even real. At the same time, it results from practical skills. A person who understood how to chip away flakes to create a delicate hand, to turn a crack into a pair of legs, or to dot a little face surely had learned these things by mastering the craft of making the late Acheulean tool kit.

So what do we have here? Not a work of art, but a craft product that identifies a topic just for the whimsy of it. The artisan had to be someone who noticed and paid attention to resemblances, and who had the skill to act on that recognition. This interest in resemblances is new. The "Venus" is the first archaeological find that suggests the presence of an interest in the resemblances that make metaphor possible. The artisan who shaped this "Venus" had created a physical metaphor, taking one thing, a stone, and turning it into another, a person.

Although not every expert is persuaded that the Tan-Tan "Venus" marks an important turning point, the archaeological finds for the two-hundred-thousand-year period between the "Venus" (four hundred thousand years ago) and the appearance of our species, *Homo sapiens* (two hundred thousand years ago), proves that that slice of history was a surprisingly productive one. A group of wooden spears carved from spruce trees was found in a coal mine in Schöningen, Germany. Radiocarbon dating puts them in a narrow range from 380 thousand to four hundred thousand years old. Amazing as this finding was, it confirmed an earlier discovery. In 1911 a lance tip made of wood and dating from the same period as the Schöningen find was uncovered in Clacton-on-Sea, England. Wooden artifacts rarely survive a few centuries, so the survival of a set of artifacts for hundreds of millennia implies that wooden tools were already a commonplace at that time.

Evidence of minerals ground up to produce coloring (ochre) and dated to about three hundred thousand years ago was found in the Twin Rivers cave near modern Lusaka, capital of Zambia in central Africa. The usual reason for grinding ochre is to make paints. There are no rock paintings associated with the cave, but ochre is frequently used as body paint. If the body paint was symbolic, the age of mythology had already begun, but we have no supporting evidence for that claim. More likely, the Twin Rivers people were living in an age of metaphor that transformed objects (chiefly themselves) into other things.

Another common use for ochre is in folk medicines, or to cure hides, or to attach wooden handles to stone blades. Thus, we cannot insist on any particular explanation for how the

ochre was used, although it is apparent that whatever use was
made of it was imaginative, and that its makers were capable
of finding uses that were nowhere nearly as obvious as the result of knocking a cutting flake from a piece of stone.

Also dating from about three hundred thousand years ago
and persisting until late in the last ice age were cutting blades
made in the Levallois style. Like other cutting blades made with
the help of tools, Levallois blades were created by knocking
flakes off from a core stone. But Levallois cores were prepared
in advance so that as blades were chipped off they were sharp
on all edges. This style is primarily associated with the human
lineage, but at the same time a similar style using prepared
cores, the Mousterian, appeared in the Neanderthal lineage.
Also found in Twin Rivers cave, this time from about 250 thousand years ago, are small blades that seem to require handles.

It appears beyond question that the human lineage of the
period immediately before *Homo sapiens* already had a complex culture and numerous practical traditions that required
rich verbal support. Yet there are still scholars who insist that
the earliest language appeared only about one hundred thousand years ago. They base their argument on the principle that
language depends absolutely on symbols and that a "human
revolution" based on symbols occurred one hundred millennia ago. Two American anthropologists, Sally McBrearty and
Alison Brooks, wrote a classic rebuttal to that position in a
paper titled "The Revolution That Wasn't." The paper was
unusually long, covering over one hundred pages and proposing what this history has shown: Starting in deep time, there
was a slow transformation of the lineage into full, recognizable
human beings.

There has, however, been a revolution in our understanding of human origins in general and language origins in particular. We used to think of the appearance of *Homo sapiens* as a beginning; the start of humanity, the birth of speech, and the onset of culture; now we see our species as the climax of a biological and cultural metamorphosis that was as slow as it was thorough. Revolutions can accomplish great surface changes, but evolution changes matters down to the marrow.

Biologically, the change in the human brain brought up the rear, following the erect gait and furless body, but it produced the most novel result—an organism that has to learn the fundamentals of living with its own kind. The *Homo* brain began growing as a scaled-up ape model, and it gradually took on some novel features—such as the ability to combine topics with news, and to link reflexive and deliberate attention—so that it eventually became a distinctly human brain. And even then it kept on growing. Human brains make many cross-sensory associations. Since the time of *Homo erectus*, there has been a dramatic expansion of a region of the front lobe, creating circuits that combine different sensory modes (e.g., sounds and sights). These cross-sensory associations allow for greater intellectual flexibility than is available to apes, and they permit more abstract association. Although the human brain did a little more growing beyond the two-hundred-thousand-years-ago mark, the human brain was essentially in place at the start of *Homo sapiens*.

We have already seen that *Homo sapiens* began with a rich linguistic heritage that included the ability to speak full sentences, make use of resemblances, and take advantage of specialized vocabularies. When it comes to abstract reasoning, their language was not ours, but neither are most of the known languages and civilizations. Until quite recently, most cultures expressed abstractions mythologically, as the doings of gods or spirits. And even today mythological and magical speech probably outnumbers technical and scientific remarks. Once metaphor had appeared, generalization was always a biological possibility, although culture still weighed heavily against it.

More than a million and a half years before *Homo sapiens*, language served as a bridge between biology and culture. It is anchored to both sides of that divide and depends on both anchor points. You can call speech biological or cultural, according to your concerns and needs of the moment, and you will not be wrong unless you insist that your classification is the whole story. By the start of our species, the bridge permitted extensive unloading of biological baggage, transferring so much to the cultural side that human newborns need to learn how to be who they are. Animals just know, but human identity depends mightily on who else is present. If we are law-abiding, it is because we live in a law-giving group. If we are storytellers, it is because we live among language users.

By the time of *Homo sapiens*, this transfer of controls from biology to culture was so extensive that even by the onset of puberty, humans still had not learned all they needed to know. Apparently one of the important differences between us and Neanderthals is that we mature at a much slower rate. Neanderthal ages were originally estimated by comparing their skeletons with human skeletons. Thus if a five-year-old human boy would have a skeleton like one found in a Neanderthal, anthropologists tended to say they had found a Neanderthal who died at age five. But using recent imaging techniques to study teeth, anthropologists have obtained much more accurate data. Teeth grow like trees, leaving a new growth ring every day. Counting them reveals that the five-year-old Neanderthal was really only three. The seeming teenage Neanderthal was a preteen. Neanderthals, in other words, grew at an ape's pace. *Homo sapiens* has slowed everything down so that when we reach puberty we still have not attained our full skeletal size.

The break with ape growth gives us adolescence, a remarkable period in which individuals are sexually mature but still dependent on adults. It is also a time when humans become locked into a culture. At about age twelve they lose the ability to learn a new language without an accent. As Dictionary Johnson put it, their tongues become stiff. Adolescence is also a time when many cultures have coming-of-age rituals and youth embark on apprenticeships in the ways expected of them.

This prolonged growth is perhaps what separates *Homo sapiens* from its ancestors in the human lineage and its kin lineage of Neanderthals. Each *sapiens* generation is born with very little knowledge of what to do. They know to suck, to vocalize, and to crawl before they walk. Compared with the rest of the zoological world, however, they are bundles of unfocused potential. By the time they stop growing, seventeen or eighteen years later, they are on a path, part of a culture, speaking one way and not another. Culture itself can often seem like a kind of enforced way of seeing and being—eyeglasses that show you the world according to its prescription. But brave observers can change the prescription, as they have done now for many thousands of generations.

A male and female *Homo sapiens* stand together near Pinnacle Point cave in southern Africa, near where Cape Town now stands. Both are naked, but they might have denied that statement, for their bodies and faces are streaked with red ochre. A bird twitters in a tree and the pair looks up at it. The scene is set 164 thousand years ago—approaching 468 thousand generations after the founding of the human lineage— and the couple invent some brief story about what the bird is saying. Standing at Africa's southernmost point, the world lies all before them and choices are scattered everywhere. The weight of over two and a half million years of *Homo* ancestors and experience has guided them to this point, but they are humans now and will have to find their own way.

EPILOG: ANOTHER CONFESSION

I N THE AUTUMN of 1953, I was eleven years old and chanced to wonder about the start of language. At the time, I was reading a comic book that included something about the cave dwellers of Lascaux, so I imagined a meeting where cavemen got together to agree on the words for various things. Young as I was, it did not take me long to realize that, without language already existing, no such meeting was possible. *Ha*, thought I, and I never forgot that circular puzzle. Language depends on learning, but learning depends on language. A lifetime later, I am satisfied that I have at last seen the outlines of an exit from that circle.

We started vocalizing for reasons having nothing to do with speech; later, sounds were given a pointing role. This process of exaptation, or pre-adaptation, adding functions to a biological trait, is familiar. A classic example is the evolution of the feather. It seems to have appeared first as a means of insulation, the very use humans still have for feathers when we make quilts. Later, feathers proved useful in flight as well. So it turns out that the puzzle that first bothered me when I was eleven was not really the most challenging part of accounting for language origins. Breaking into circular logic is just a day's work in the story of evolution. Much more difficult has been the issue of determining what had to evolve, what was already present at least potentially, and what we had to be smart enough to invent. There are still fierce debates on all these points.

At least we now debate the issues. In 1953, however, nothing was debated. Not only was nothing known about language origins, it was disreputable to raise the question. Ten years after I first noticed the origins question, during my college days, I attended a discussion session with a visiting literary critic, a Jesuit priest named Walter Ong. At some point in his presentation I asked how language could even have begun. He told me, "That is the forbidden question," and went on to say that the Paris Linguistic Society had officially banned the question in 1860. The London society followed suit a few years later.

Nonetheless, I did continue to wonder, and by 1970 I had realized that there was probably an evolutionary side to the story of language origins. I hunted up and read Eric Lenneberg's 1967 book *The Biological Foundations of Language*. Since that reading, I have had no doubts that speech has a biological and evolutionary basis. Twenty more years had to pass, however, before the subject gained respectability. In 1990, two MIT psychologists, Steven Pinker and Paul Bloom, published a paper titled "Natural Language and Natural Selection." It said there had to be an evolutionary account of language origins, and that admission started a gold rush of further study. The sudden flood of ideas and interest proved that I had not been alone in wondering about where words came from.

A conference on language evolution held in Edinburgh in 1996 was so successful that a follow-up conference has been held every two years. By now, besides the big conference, there is a steady stream of workshops, seminars, and meetings on this or that aspect of language origins. A mass of data has been assembled. Father Ong died in 2003, and by that time enough progress had been made that, I like to think, he might have said to himself, "I should never have tried to discourage that young fellow."

And a few years later I returned to the subject that I had never forgotten. It struck me that time was sweeping by and that after half a century I should make a serious effort to understand speech origins. Having survived into the twenty-first century, I did a new century deed and started a blog with the same title as this book. I began the first post on my blog by writing a sentence that made me wince at my own audacity. I said that Babel's Dawn "aims to become the main source of news and information about the evolution of speech." Was there even such a thing as news about speech origins? It turned out that, yes, there was a steady flow of papers to discuss and conferences to cover. It also became apparent that there are heavy quarrels over just about each detail of the story.

With so much in dispute, readers have a perfect right to ask how much of this book they should take seriously. Does it have any more legitimacy than Rudyard Kipling's *Just So Story* about how the leopard got its spots?

Jeremy Freese, now chairman of the sociology department at Northwestern University, once gave an answer that has encouraged me to keep plugging. He told a group that "evolutionary social science has two distinct projects," one historical and one deductive. The historical project in this slender volume has taken the observable facts of language and human evolution

and assembled them into an account of the emergence of a talking lineage. The deductive project "involves taking knowledge
of [language] and about the logic of the selection processes and
using that to develop new verifiable insights" into language. In
this book the deductive project has looked at what the origins
tell us about both language and its users.

"Evolutionary theorizing of this sort," Freese continued,
"is regularly criticized as storytelling, and storytelling it often
is—but the narrative history of our species is a story, and the
stories we tell may be better or worse approximations of this
real story, for which consistency with available information is
our guide."

We have more facts available than a naïve skeptic might
suppose. Evidence for speech origins comes from the study of
language acquisition in children, comparative data where animal and human behavior converges, investigations of Creole
languages, experiments on artificial language learning, medical
research into linguistically impaired individuals, and computer
simulations that calculate the implications of our present understanding. We also have a mountain of data about human
origins in general. All this information allows for a set of hypotheses about what happened, when it happened, and what
it happened to. These hypotheses form the story told in this
book. What would I say to any creationists who object that it
is all "mere hypothesis"? Grant the point, but add as well that
it is based on evidence and will be modified by better evidence,
not by dogma.

Together, this material gives us a two-layered picture.
One layer displays the changes in the human lineage during
the past six million years. The other shows the development

of language. There is good knowledge for creating both the species-origin layer of this portrait and the speech-development layer. The major problem has been getting the two in register so they combine to form a sharp image. I have tried to be conservative, giving as late a date as I believed the story could sustain. The first word may very well predate 1.8 million years ago, the first full sentence that combined a subject and a predicate might have been before 1.5 million years ago, episodic accounts may be older than nine hundred thousand years, and metaphors may be more than three hundred or four hundred thousand years old. I would be surprised if the dates are premature, although I concede that my capacity for surprise does not count as evidence. A scenario like this one may read more like literature than science, but it is a serious hypothesis about what happened.

More important is testing the deductive part. Are the conclusions that emerge from the history correct? Useful? I did not realize when I was eleven that a question about the origins of anything meant asking about that thing's essence. Pinker and Bloom were older than eleven when they published their groundbreaking insistence that language had evolved, but they did not seem to expect that the work they were calling for might change our understanding of language. They suggested that language's function was the obvious one of "communication." That, however, was at the start of the explosion of inquiry. There are many more ideas now.

Ideas about human origins have changed plenty since I first began wondering about language. In those days, the human lineage was thought to have arisen in Asia and our big brains were assumed to be older than upright walking. Meanwhile, apes were supposed to have no tools, no gestures, no capacity for reasoning, and no ability to learn any form of language. These errors reflected a larger truth: We still had no idea what distinguished humans from apes.

By a stroke of good luck, I once attended a luncheon where that ignorance became instantly apparent. In 1964, I was home from college and my father took me to a luncheon at the National Press Club where the speaker was the legendary paleontologist Louis Leakey. Earlier that day, Leakey had announced the discovery of *Homo habilis* and now he was speaking to the press less formally. As part of his presentation he told us about Jane Goodall's recent discovery that chimpanzees in the wild were making tools to catch ants. It was time, he told the stunned room, to come up with a new definition of being human.

Language has now become a central element in the modern approach to understanding humanity, and many themes linking language and humanity have become apparent in the literature of language origins. As I worked on my blog and read a snowfall of academic papers, it was the themes that kept me going, and which made me sympathize with researchers whose guiding theories were not my own. For example, a brief paper that appeared in 2005 included many of the themes encountered in our story even though the authors are much less biologically oriented than I. The paper was by David Leavens, William Hopkins, and Kim Bard with the entertaining title of "Understanding the Point of Chimpanzee Pointing," and it showed a common set of facts leads to common themes. Their presentation included matters that run through this book: the importance of topics, mutual attention, the speech triangle, and the way human society brings out certain traits that, in principle, were already available in apes.

The Leavens team's paper introduces two of those themes in its opening sentence, "A defining characteristic of the human species is our capacity to rapidly establish topics for mutual contemplation."

Animals by their nature are interested in at least some things that stand outside themselves. Any visitor to a national park that boasts ample wildlife is likely to have seen animals come to attention when the visitors arrive. So I make no claim that only humans are interested in topics. Humans, however, are interested in more of them, probably because we can find ways to turn seemingly neutral topics to our advantage. That trait appeared in our history at least as early as the emergence of Oldowan technology, 2.8 million years ago, in which one tool was used to fashion a second tool. Even a topic as abstract and nerdish as speech origins has its practical payoff. Thanks to this lifelong project I have a sharper understanding of what I am doing when I write.

Equally important in the team's first sentence is "mutual contemplation." People are unusual in their joint contemplation of topics. We also contemplate singly, but it is mutual contemplation that distinguishes us from the giraffes and other animals staring out at the world and responding to what they perceive. Mutual contemplation is essentially unknown in other species, but is part of every human group. All human societies have wise counselors, individuals who have been granted importance by the group because of the value of their words rather than because of their physical dominance. Some scholars, notably Jean-Louis Dessalles, have suggested that the reputation and honor was *the* reason we talk. That may be an overstatement, but there is no getting around the benefits that come to individuals and the group from jointly contemplating some topic.

A third theme that my blog work emphasized comes at the end of the team's first paragraph. It refers to "a referential triangle that incorporates distant objects into the relationship between a signaler and the recipient of the gesture." They are talking about pointing, but I trust readers will recognize here what I have been calling the speech triangle: the joint attention of a speaker and listener focused on a topic.

The team's paper does reflect one old assumption, that pointing and speaking are functionally different activities. The triangular relationship common to both actions is ignored. The Leavens team's paper denigrates pointing by saying it lacks meaning, or as cognitive psychologists say, lacks representations. Pointing, the paper says, does "not stand for or represent the objects indicated in the way that words do . . . words that represent particular entities are not iconic—e.g., the relationship between the physical features of the word *dog* and actual dogs is arbitrary." These authors are saying nothing controversial here, but their work reflects none of the rising interest in the relation between attention and language.

The child psychologist Jerome Bruner began writing about joint attention and human development as early as 1975. By 1995 the autism community was taking joint attention seriously too. In the final years of the last century there was a sudden eruption of papers examining the role of joint attention and language development in children, and that work accelerated still further during the first five years of the new

century, with students of language and language origins joining the chatter.

As described by Leavens' team, the referential triangle
presents an objective triplet, a "relationship" between two individuals and an object. The speech triangle invoked throughout this book has been a subjective relationship, dependent
upon joint attention. It turns out that when you consider the
subjective triplet instead of the objective one, the problem of
speech origins becomes much simpler.

First off, attention gets rid of the problem of evolving a
representation system. Instead of a word representing something, it directs your attention to something. The difference
from an evolutionary standpoint is quite simple. Attention and
shifts in attention are very old, so old that this history can take
them for granted. It is a good thing too that we can grandfather
them into the history, because we have no idea how attention,
sensation, or awareness works. We do know, however, that
the link between a word and a perceptible thing can be learned
by association, which is why dogs can learn their names. So
we do not need to postulate an intermediate representation between a word and its association.

Representations only become necessary when words point
to something imperceptible. Instead of being associated with
something concrete, imperceptible words point to definitions.
For example, learning Euclidean geometry means learning a set
of mutually dependent definitions and rules, not associations.
These require a formal education that teaches the definitions
and how to use them.

One of the great mysteries of language is the way children pick it up speedily and without dedicated training. It is

an impressive feat, made all the more so if you suppose they are learning something like geometry, a system of interlocking definitions and fixed rules. If they learn the concrete side of the process first, however, their accomplishment is admirable, but not contrary to everything else we know about education.

These days, mathematicians, lawyers, and scientists use language to establish imperceptible relationships, but initially such relationships may have emerged through ceremonies or rituals rather than by speechifying. As Terrence Deacon stated in his fine book on language origins and symbols, "it wouldn't be surprising to find that the earliest social symbols for establishing reproductive roles were created in a ritual."

Attention's antiquity brings us to a final theme discoverable in the Leavens team's paper: Speech depends on some capacities that apes already have. The paper's thesis is that although chimpanzees do not point in the wild, captive chimpanzees will point spontaneously to food that they need a human to supply. Thus, the authors conclude, no special evolution was required to begin pointing. The genetic capacity was already there once there was a reason to support the pointing behavior. I have no trouble with that idea. The fewer ducks evolution had to set in a row just to get started, the more I like it.

If we take these themes together—interest in neutral topics, mutual attention, the speech triangle, and latent abilities—we can see how they suggest a solution to the reconsideration of human nature that Leakey called for. Humans are born with surely the fewest environmental adaptations of any mammal that has ever lived, but we end up with a series of artificial adaptations that make us the dominant species in whatever habitat we happen to occupy. How are we so successful? The adaptations we are born with are not even adaptations to a habitat, they are adaptations to living in a community. These adaptations reflect the themes we have been looking at.

We can continue with the pointing example. Apes are smart enough and have enough control over their hands to point. They lack both the motivation for cooperative pointing and the willingness to trust another's pointing as helpful. Human infants are moved to help and willing to trust at an early age. These adaptations to community life are part of being human.

Another human adaptation is an interest in seemingly useless things. Interest in neutral topics leads to the discovery of how to adapt to the habitat. It is a capacity humans desperately need while other animals who are already well adapted to their habitat do not need at all.

Still a third human adaptation is a taste for joint attention. Like pointing, it is possible to get captive apes, especially young ones, to pay attention to something a human attends to. What is different in the human is the sense of solidarity that makes a person join in with another, and tolerance to attend to something that does not grab their immediate interest.

Each of these traits supports speech, but they are not the usual traits listed as prerequisites. Instead of mechanical abilities such as the capacity to merge language elements appropriately, or to engage in symbolic thinking, this account rests on helpfulness, trust, interest in things of no obvious value, solidarity, and tolerance within the group. In a survey of languages since the dawn of writing, Nicholas Ostler found that in all the spread and contraction of languages, "a language does not grow through the assertion of power, but through the creation of a larger human community." Once we started cooperating, we liberated a whole series of capacities that apes have but do not use.

Western social philosophy, especially in the United States, emphasizes individualism, but when you look at how democratic nations are organized—with business, social, religious, and cultural groupings everywhere—it is clear that it does not mean the kind of lonely sink-or-swim competition that dominates the rest of the primate world. Chimpanzees may hug one another and come together in large groups to spend the night, but they are in an all-against-all struggle to pass their genes along. Humans are in a struggle to keep their groups alive, and after nearly three million years of *Homo* cooperation, the number of groups that put demands on our loyalty has become remarkably complicated.

Unexpected as such a view of humanity may have been— and I can tell you that I was not expecting it when I began my study—it has the strength of answering the question of why only the human lineage evolved language. As the only communal primates, we are the only primates who need it.

Freese's two projects come together. The deductive part supports the history part, and vice versa. The history leads to the deduction that language is a cooperative system for directing joint attention to topics and their details. Other animals have no such system because they have no use for cooperatively considering topics, while humans need the process so desperately that they evolved all the instincts for sharing, imitating, and interacting cooperatively that make speech inevitable. Language is the keystone that holds understanding and cooperation together.

À Word About Those Names

I

N ORDER TO individualize these unnamed ancestors, I have given them particular names, and since they are the ancestors of everybody alive in the world today, I have used names from all the world's cultures and civilizations. Below is a list of the names and their origin:

Alisa (RUSSIAN)

Anne (ENGLISH)

Capac (PERU)

Dakila (TAGALOG)

Gärd (SWEDISH)

Harun (TURKISH)

Iskuhi (ARMENIAN)

Ling (CHINESE)

Mahasti (IRANIAN)

Ng'ula (BANTU: VIDUNDA)

Olatunde (YORUBA)

Sara (HEBREW)

Shoichi (JAPANESE)

Taranga (MAORI)

Yikin (MAYAN)

Notes

S OME OF THESE references end with a date. These indicate the day of posting on the Babel's Dawn blog where the source and matter was discussed, usually more fully.

ENTRANCE HALL

3   *a million children could not invent a language*, (Boswell, 1993, p. 1080); October 26, 2006.

4   *Creoles were created by the children of those pidgin-speaking adults*, (Bickerton, 1985).

5   *the children create a sign language*, (Kegl, Senghas, and Coppola, 1999); November 7, 2006.

5   *hunger for language*, (Kegl, 2006); November 7, 2006.

7   *particularly biologists*, (Bichakjian, 2006); November 8, 2006.

10   *The simplest account of the "Great Leap Forward" in the evolution*, (Chomsky, 2004); February 24, 2008.

12   *In most disciplines*, (Kenneally, 2007, pp. 200, 264).

14   *The speech triangle summarizes the community structure*, (Leavens, Hopkins, and Bard, 2005); January 1, 2007.

14   *slap the ground*, (Tomasello, 2008; Dawkins and Krebs, 1978, p. 73); March 14, 2010.

15   *a whimpering chimpanzee child*, (Tomasello, 2008, p. 5).

15   *There is a straightforward, Darwinian explanation*, September 28, 2008.

GALLERY I

Sara

17 *she is her own distinct species*, (Crompton and Thorpe, 2007).

18 *eight studies of nuclear DNA*, (Chen and Li, 2001, p. 452; Eizirik, 2004, figure 4; Glazko and Nei, 2003, p. 432; Page and Goodman, 2001, table 4; Satta, 2004, p. 486; Stauffer, 2001, p. 469; Steiper, 2004, table 2; Wildeman, 2003, p. 7185).

18 *arrived at a date from a probabilistic model*, (Wilkinson, Steiper, Soligo, et al., 2011).

19 *According to a study published in 2002*, (Britten, 2002).

19 *mice match us 85*, (Rult, 2007).

19 *duckbilled platypus shares 82*, (Warren et al., 2008).

19 *Ah, so!*, (Jablonka and Lamb, 2005, p. 209); November 30, 2006.

20 *vocalizations are a common part of voluntary gestures from apes*, (Pika, 2008); May 4, 2008.

21 *reaches into her child's mouth*, (Knight, 2002, p. 143); November 6, 2006.

22 *vervets take context into account when they respond to an alarm*, (Dessalles, 2007); February 28, 2007.

22 *tack sounds to the front or end of a signal*, (Endress, Cahill, Block, Watumull, and Hauser, 2009); July 12, 2009.

23 *the ape listener is not anywhere nearly so strongly constrained*, (Seyfarth and Cheney, 2010); March 7, 2010.

23 *Apes are very capable at perceiving, navigating, and drawing logical conclusions*, (Zuidema and Verhagen, 2010); March 21, 2010.

25 *a kind of activity*, (Marchetti, 2006); October 29, 2006.

26 *They recognize the two kinds of attention*, (Mundy and Newell, 2007); February 17, 2008.

27 *how evolution can produce a distinctly different creature*, (Margoliash and Nusbaum, 2009); November 22, 2009.

Harun

31 *evolutionary stable strategy*, (Maynard-Smith, 1979; 1982).

34 *cases of "adoption"* (Boesch, Bolé, Eckhardt, and Bosesh, 2010).

Alisa

38   *they are unusual or unexpected*, (Dessalles, 2007, p. 27); February 27, 2007.

39   *Starting 4.7 million years ago, and persisting* (Trauth, Maslin, Deino, et al., 2007).

39   *had begun two hundred thousand years earlier*, (ibid., figure 5).

40   *cooperation is the best strategy*, (Kazakov and Bartlett); October 9, 2006.

40   *a kind of species pump*, December 4, 2007.

40   *The oldest Ardipithecus fossil yet known*, (Trauth, Maslin, Deino, et al., 2007, figure 6).

Yikin

41   *two more complete wet/dry phases*, (ibid., figure 5).

42   *the famous Laetoli footprints*, (Leakey, 1979; Sawyer and Deak, 2007, p. 7).

43   *By Yikin's time the head was located much more directly over the spine*, (Nevell and Wood, 2008).

44   *Short legs lower a person's center of gravity*, (Carrier, 2007); March 18, 2007.

44   *louse DNA*, (Wade, 2007); March 11, 2007.

46   *a substitute for grooming*, (Dunbar, 1997).

47   *"cheesecake" effect*, (Levitin, 2006).

47   *still had laryngeal air sacs*, September 20, 2006.

48   *Ape calls are*, (Zuidema and de Boer, 2009); May 9, 2010.

49   *put the baby on the ground*, (Falk, 2009).

50   *like a wildebeest*, (I've seen it myself, Serengeti Plains, February 1980).

Anne

51   *still another wet/dry phase (the fourth of the series) began*, (Trauth, Maslin, Deino, et. al., 2007, figure 6).

54   *apes make voluntary, illustrative gestures*, (Tomasello, 2008).

54   *learn signing a month or so before they can speak*, (Acredolo, Goldwyn, Abrams, 2009).

54   *Illustrative gesture, of course, is also a normal part of* speech, (see e.g. Goldin-Meadow, 2008).

55   *dorsal auditory pathway*, (Aboitiz and Garcia, 2009, p. 71); July 5, 2009.

55   *You cannot tell an animal anything*, (Tomasello, 2008).

55   *maintain contact with one another as they fly*, (Graber and Cochran, 1959).

55   *lung power instead of the subtleties of vocal-tract control* (Riede, Fisher, and Goller, 2010).

56   *A Japanese primatologist*, (Masataka, 2009); August 9, 2009.

56   *sign language rested on four inherent qualities*, (Kegl, 2006); November 7, 2006.

56   *melody arc*, (Falk, 2009, p. 83); March 29, 2009.

Mahasti

58   *a primate mother has clung to her dead newborn*, (I've seen it myself, Aberdare National Park, Kenya, November 1977).

59   *"tools" are really weapons*, (Ardrey, 1961).

60   *a chimpanzee in a Swedish zoo*, (Sample, 2009).

60   *Anyone watching the action*, (Kevinropelato, 2008).

60   *wrist action required for effective throwing*, (Ambroose, 2001, p. 1749; Thorpe, Crompton, et al., 1999; Young, 2003).

63   *The previous wet/dry phase*, (Trauth, Maslin, Deino, et. al., 2007, figure 6).

Ling

65   *A mother zebra or wildebeest* (I've seen it myself with both species, Serengeti National Park, Tanzania, calving week, February 1980).

65   *a dramatic change in habitat*, (Trauth, Maslin, Deino, et al., 2007, figure 6; based on Levin, Simpson, Semaw, and Rogers, 2004; and Wynn, 2004).

69   *multilevel evolutionary theory has made a striking comeback*, (the basic material for this chapter comes from Wilson and Wilson, "Rethinking the Theoretical Foundation of Sociobiology," 2007; Wilson, D. S., *Darwin for Everyone*, 2007; Haidt, 2007; Fletecher, Zwick, Doebeli, and Wilson, 2006; Borrello, 2005; Goodnight, 2005; Nunney, 1999).

69   *selfish-gene theory . . . was given a mathematically solid foundation in a series of papers*, (Hamilton, 1963; Hamilton, 1964; Wilson, E. O., 1975; Dawkins, 1976).

70   *look in the opposite direction. Perhaps speech arose as a tool for deceiving our neighbors*, (for some discussions of lying see Zahavi and Zahavi, 1997; also Dessalles, 2000).

71   *understand human speech in terms of selfish genes*, (Dessalles, 2007); March 29, 2007.

71   *something of a relief in 2005 when E. O. Wilson surprised the world of insect specialists*, (Wilson and Hölldobler, 2005).

72   *E. O. Wilson first coined*, (Wilson, E. O., 1975).

72   *an elegant theoretical foundation*, (Wilson and Wilson, 2007, p. 339).

76   *the honeybee genome*, (Weinstock, Robinson, et al., 2006); October 26, 2006.

76   *Pseudomonas fluorescens*, (Wilson and Wilson, 2007).

76   *Spinner dolphins have been seen hunting together*, October 8, 2008.

77   *takes almost ten times as much energy*, (Isler and Van Schaik, 2009, p. 392).

78   *each lineage faces a 'grey ceiling,'* (Isler and Van Schaik, 2008, abstract).

78   *Cooperative child rearing can reverse the findings*, (Isler and Van Schaik, 2009, p. 395).

78   *speech is a "fundamentally communal" behavior*, (Wilson and Wilson, 2007).

78   *a shift in the balance between levels of selection before [it] could evolve*, (ibid., p. 343).

78   *The other female is . . . ready to help with a delivery*, (Walrath, 2006).

79   *the chimpanzee Y chromosome is radically different from the human version*, (Hughes, Skaletsky, Pyntikova, et al., 2010).

80   *wholesale renovation and remodeling*, (Bagley, 2010).

80   *allow males to act cooperatively*, (Deacon, 1997).

80   *females had to rebel against the males*, (Knight, 2002).

Ng'ula

81   *the weather had dried up*, (Trauth, Maslin, Deino, et. al., 2007, figure 6).

81   *the whites of the eyes*, (de Waal, 1982).

82   *mobsters wearing sunglasses*, (Knight, 2002, p. 144).

85   *embedding one clause in another*, (Fitch, 2010, p. 100).

85   *Noam Chomsky, has even speculated*, (Hauser, Chomsky, and Fitch, 2002).

86   *the royal road to language*, (Butterworth, 2003).

Shoichi

88   *spread as far as Indonesia*, (Lewin, Curtis, and Swisher, 2001).

89   *the borderline between colors*, (Puglisi, Baronchelli, and Loreto, 2008); June 22, 2008.

91   *at least fully functional*, (Bickerton, D., 2009).

93   *All signaling systems must begin*, (Nobel, de Ruiter, and Arnold, 2010, p. 73).

CENTRAL HALL

96   *there must not be any difference in kind,* (Penny, 2010); July 11, 2010.

100   *B. F. Skinner, attempted to prove,* (Skinner, 1957).

101   *Noam Chomsky attacked its theory from many angles,* (Chomsky, 1959).

107   *functions other than animal communications,* (Deacon, 2010); April 25,
      2010.

108   *not generated in special brain areas,* (Pulvermuller and Assadollahi, 2007;
      Pulvermuller, 2010); March 11, 2008.

108   *how to make different Stone Age tools,* (Stout, Toth, Schick, and Chaminade,
      2008); June 8, 2008.

109   *cost in choking is serious,* March 26, 2010.

110   *the circuit provides a shortcut,* (Aboitiz and Garcia, 2009); July 5 and August
      2, 2010.

111   *dorsal auditory pathway,* (ibid., p. 71).

111   *an unprecedented richness,* (ibid., p. 79).

114   *know what lies beyond their skins,* (Hurford, 2007); October 21, 2007.

114   *depends on effective social coordination and interaction,* (Tylen, Weed,
      Wallentin, Roepstorff, and Frith, 2010); January 31, 2010.

114   *Joint attention leads to joint action,* (Gardenfors, 2004); April 7, 2010.

117   *Biological changes must get the science right,* (Chater, Reali, and
      Christiansen, 2009); February 1, 2009.

GALLERY II

Iskuhi

121   *toddlers have been heard in bed at night practicing words,* (Bolles, 1982).

122   *contributes to vocal learning in hummingbirds, parrots, and even some bats,*
      (Fitch, Huber, and Bugnyar, 2010); April 11, 2010.

123   *have nothing like the expressiveness of dogs,* (*Nova,* 2010); November 14,
      2010.

123   *Today we have a list of over one hundred genes,* (Konopka and Bomar,
      2009); November 15, 2009.

124   *belongs to a broader tradition,* (e.g., Levitin, 2006); September 21, 2006.

125   *capable of making as many as ten different morphemes,* (de Boer, Modelling,
      2010); October 18, 2010.

126   *biology plays a role,* (Stromswold, 2007); March 4, 2007.

126   *path that reflects their perceptions*, (Oudeyer and Kaplan, 2007); August 26, 2007.

126   *speech that survives is user-friendly*, (ibid.).

130   *asking a question simply assumes many elements of a speech triangle*, (Tomasello, Carpenter, and Liszkowski, 2007); June 4, 2007.

130   *grew by about a tablespoon per hundred thousand years*, (quoted in "Becoming Human—Special Edition," 2010).

131   *all reflected one toolmaking tradition*, (Stout, Semaw, Rodgers, and Cauche, 2010).

131   *buried in the mountains of southwest Asia*, (Pontzer, Rolian, Rightmire, et al., 2010).

132   *slower the process of taking on a modernized form*, (Lieberman, Michel, Jackson, Tang, and Nowak, 2007); July 20, 2008.

132   *heavy gravity for resisting any change*, (Feher, Wang, et al., 2009); May 10, 2009.

133   *a few telling exceptions*, (Bolles, 1982, p. 3).

133   *exactly what happens with any other common word*, (de l'Etang, Bancel, and Bengtson, 2006; Bancel, de l'Etang, and Ruhlen, 2006).

Gärd

142   *became genetically able to digest the lactose*, (Wray, 2007).

143   *Fluency returns last*, (Ross, E. D., 2010, p. 4); May 23, 2010.

144   *consisted of mostly news*, (Hurford, 2009); October 4, 2009.

145   *Chimpanzees are smart enough to imagine a desirable outcome*, (Kenneally, 2007); July 22, 2007.

145   *there are two processes here*, (Falk, 2009, p. 84); March 29, 2009.

146   *have more nerves than any other muscles in the human body*, (Karpf, 2006, p. 27); November 5, 2006.

Dakila

150   *a taste for mirroring*, (Kegl, 2006); November 7, 2006.

151   *chimpanzees can learn an arbitrary four-action fixed sequence*, (Whiten, 1998).

151   *leads to "improvement" in complex language production*, (Conway and Christiansen, 2001, p. 540).

154   *A film clip on a monitor*, (Goldin-Meadow, 2008); March 30, 2008.

155   *It is not merely that a word like she is symbolic*, (Bickerton, D., 2008);
      March 30, 2008.

Capac

159   *Dediu and Ladd's pioneering work*, (Dediu & Ladd, 2007); September
      30, 2007.

161   *verbs as a class appear only in the third year*, (Brandone, Pence, Golinkoff,
      and Hirsh-Pasek, 2007); August 12, 2007.

163   *when an angry elephant backs* away, (I've seen all these things myself in
      South Luangwa National Park, Zambia, 1978).

165   *A bird's eye view*, (Tomasello, 2008); September 28, 2008.

Taranga

169   *A natural behavior whose expression has been prevented*, (Dessalles, 2007,
      p. 75); March 6, 2007.

Olatunde

170   *time of the mid-Pleistocene revolution*, (Maasch and Saltzman, 1990).

175   *Serial verbs*, (Bickerton, D., 2008); April 20, 2008.

176   *she had embarked, sailed, landed, explored*, (James, 1902, book III, chapter
      1).

176   *An episode joins a set of facts or associations in time*, (Corballis, 2009);
      March 1, 2009.

177   *as much difficulty imagining new experiences*, (ibid.).

177   *ground a sentence's meaningful words*, (Deacon, 2009); September 21, 2009.

178   *Language translates all non-visual relationships into spatial relationships*,
      (quoted in Dessalles, 2007, p. 239); March 26, 2007.

178   *Syntax does the work of perception*, (Arsenijevic, 2008); May 28, 2008.

GRAND HALL

182   *the leading textbook on language evolution*, (Fitch, 2010).

182   *called* The Language Instinct, (Pinker, 1994).

183   *what would syntacticians have to theorize about*, (Hurford, 2009).

184   *April Benasich, reports*, (Benasich, 2007); May 6, 2007.

185   *Gary Marcus has suggested*, (Marcus, 2008); June 2, 2008.

187   *cannot account for content*, (Reboul, 2010, p. 435); June 20, 2010.

188 *rejects Darwin's theory*, (Fodor and Piatelli-Palmarini, 2010); February 14, 2010.

188 *word meanings are represented in mind as assemblies of basic concepts*, (Pinker, 2007, p. 91;) September 9, 2007.

189 *the mystery of interest and selective attention*, (quoted in Richardson, 2006, p. 119); March 15, 2007.

190 *a person would not be able to think deeply enough*, (Chomsky, 2007); February 24, 2008.

191 *but we lack anything even approaching a good theory of meaning*, (Fitch, 2010, p. 285); January 17, 2010.

191 *pairs sounds with meanings*, (e.g., Bolhuis, Okanoya, and Scharff, 2010, p. 750; Posner, 1989, p. 176).

193 *The earliest physical trace of controlled fire*, (Goren-Inbar, Alperson, Kislev, et al., 2004).

194 *seems to show a division of labor*, (Alperson-Afil, Sharon, Kislev, et al., 2009); December 17, 2009.

195 *the air sacs would disappear*, (de Boer, "Air Sacs and Speech," 2008); March 14, 2008.

195 *scattered across Asia, Europe, Indonesia, and the British Isles*, (Wade, 2010); July 8, 2010.

196 *create a reality*, (Lakoff and Johnson, 1980, p. 144); November 14, 2006.

198 *severe inability to comprehend metaphors*, (Rundblad and Annaz, 2010).

198 *linguistic processing may be at fault*, (Gold, Faust, and Goldstein, 2010).

198 *metaphor comprehension appears to require a strong contribution from the right half*, (Gold and Faust, 2010).

198 *hidden deep within . . . to light*, (quoted in Ostler, 2005, p. 181); September 20, 2006.

199 *660 thousand years ago, give or take 140 thousand years*, (Green, Malaspinas, Krausse, et al., 2008); August 8, 2008.

200 *required better motor skills*, (Stout, Toth, Schick, and Chaminade, 2008); June 8, 2008.

201 *Tan-Tan*, (Balter, 2009); February 8, 2009.

203 *group of wooden spears carved from spruce trees*, (Kuwenhoven, 1997); November 7, 2006.

204 *Also dating from about three hundred thousand years ago*, (Balter, 2009); February 8, 2009.

204 *"The Revolution That Wasn't,"* (McBrearty and Brooks, 2000).

205 *greater intellectual flexibility than is available to apes*, (Ross, E. D., 2010); May 23, 2010.

EPILOG: ANOTHER CONFESSION

206   *human newborns need to learn how to be who they are*, (Ross, D., 2007); November 11, 2007.

206   *human identity depends mightily on who else is present*, (Bachman, 2009); March 15, 2009.

211   *evolutionary social science has two distinct projects*, (Freese, October 2006).

212   *Evidence for speech origins*, (Gong, Minett, and Wang, 2009); September 26, 2009.

215   *defining characteristic of the human species*, (Leavens, Hopkins, and Bard, 2005, p. 185).

216   *a referential triangle that incorporates distant objects*, (ibid.).

216   *Bruner began writing about joint attention*, (Bruner, 1975).

216   *the autism community*, (Sigman and Kasari, 1995).

218   *earliest social symbols for establishing reproductive roles*, (Deacon, 1997, p. 406).

219   *through the creation of a larger human community*, (Ostler, 2005, p. 556).

Bibliography

Aboitiz, F. and R. Garcia, "Merging of Phonological and Gestural Circuits in Early Language Evolution." *Reviews in the Neurosciences* 20 (2009): 71-84.

Acredolo, L.P.; S. Goodwin; and D. Abrams. *Baby Signs: How to Talk with Your Baby Before Your Baby Can Talk.* 3rd ed. New York: McGraw-Hill, 2009.

Alperson-Afil, N.; G. Sharon; M. Kislev; et al. "Spatial Organization of Hominin Activities at Gesher Benot Ya'aqov, Israel." *Science* 326 (2009): 1677–1680.

Ambroose, S. H. "Paleolithic Technology and Human Evolution." *Science* 291 (2001): 1748–1753.

Arbib, Michael A. "The Mirror System, Imitation, and the Evolution of Language." *Imitation in Animals and Artifacts.* Edited by C. Nehniv and K. Dautenhahn. Cambridge, MA: MIT Press, 2002: 229–280.

Ardrey, R. *African Genesis: A Personal Investigation into the Animal Origins and Nature of Man.* New York: Atheneum, 1961.

Arsenijevic, B. "From Spatial Cognition to Language." *Biolinguistics* 2 (2008), www.biolinguistics.eu/index.php/biolinguistics/article/view/34/39.

Bachman, C. S. "Becoming Conscious of the Human Group." *Group Analysis* 42 (2009): 62–79.

Bagley, K. "New Clues to Y Evolution." *The Scientist* (2010), www.the-scientist .com/blog/display/56271/ (accessed January 13, 2010).

Balter, M. "On the Origin of Art and Symbolism." *Science* 323 (2009): 709–711.

Bancel, P. J.; A. M. de l'Etang; and M. Ruhlen. "The Proto-Human Words PAPA and MAMA and the Origin of Articulate Language." Cradle of Language Conference, Stellenbosch, South Africa, November 6–10, 2006.

"Becoming Human—Special Edition." *Scientific American*, www.scientificamerican .com/special/toc.cfm?issueid=44 (accessed November 22, 2010).

Benasich, A. A. *From Birdsong to Baby Talk: Studies of Language Development.* New York: New York Academy of Sciences, 2007.

Bichakjian, B. H. "Language Evolution: A Gradual Process From Cradle to Current Times." Cradle of Language Conference. Stellenbosch, South Africa, November 6–10, 2006.

Bickerton, D. *Roots of Language*. Ann Arbor: Karoma, 1985.

———. *Language and Species*. Chicago: University of Chicago Press, 1990.

———. *Bastard Tongues: A Trail-Blazing Linguist Finds Clues to Our Common Humanity in the World's Lowliest Languages*. New York: Hill and Wang, 2008.

———. "Words Are More Human Than Syntax." Evolang Conference, Barcelona, Spain, March 15, 2008.

———. *Adam's Tongue: How Humans Made Language and Languge Made Humans*. New York: Hill & Wang, 2009.

Boesch, C.; C. Bolé ; N. Eckhardt; and H. Bosesh. "Altruism in Forest Chimpanzees: The Case of Adoption." PLoS ONE 5 (2010), doi:10.1371/journal.pone.0008901.

Bolhuis, J. J.; K. Okanoya; and C. Scharff. "Twitter Evolution: Converging Mechanisms in Birdsong and Human Speech." *Nature Reviews Neuroscience* 11 (2010), 747–759.

Bolles, E. B. *So Much to Say: How to Help Your Child Learn to Talk*. New York: St. Martin's Press, 1982.

———. *Einstein Defiant: Genius Versus Genius in the Quantum Revolution*. Washington, D.C.: Joseph Henry Press, 2004.

Borrello, M. E. "The Rise, Fall, and Resurrection of Group Selection." *Endeavour* 29 (2005): 43–47.

Boswell, J. *The Life of Samuel Johnson*. London: Everyman's Library, 1993.

Brandone, A.; K. Pence; R. M. Golinkoff; and K. Hirsh-Pasek. "Action Speaks Louder than Words: Young Children Differentially Weight Perceptual, Social, and Linguistic Cues to Learn Verbs." *Child Development* 78 (2007): 1322–1342.

Britten, R. "Divergence Between Samples of Chimpanzee and Human DNA Sequences Is 5% Counting Indels." *Proceedings of the National Academy of Sciences* 99 (2002): 13633–13635.

Bruner, J. "From Communication to Language—A Psychological Perspective." *Cognition* 3 (1975): 255–287.

Butterworth, G. "Pointing Is the Royal Road to Language for Babies." *Pointing: Where Language, Culture, and Cognition Meet*. Edited by S. Kita Hillsdale. New Jersey: Lawrence Erlbaum, 2003: 9–33.

Calvin, W.H. "Did Throwing Stones Shape Hominid Brain Evolution?" *Ethology and Sociobiology* 3 (1982): 115–124.

Carrier, D. R. "The Short Legs of Great Apes: Evidence for Aggressive Behavior in Australopiths." *Evolution* 61 (2007): 596–605.

Chater, N.; F. Reali; and M. H. Christiansen. "Language Acquisition Meets Language Evolution." *Cognitive Science* 34 (2009): 1–27.

———. "Restrictions on Biological Adaptation in Language Evolution." *PNAS* (2009): 1015–1020.

Chen, F. C. and W. Li. "Genetic Divergences Between Humans and Other Hominnoids and the Effective Population Size of the Common Ancestor of Humans and Chimpanzees." *American Journal of Human Genetics* 68 (2001): 444–456.

Chomsky, Noam. *Syntactic Structures*. The Hague: Mouton, 1957.

———. "A Review of B. F. Skinner's *Verbal Behavior*." *Language* 35 (1959): 26–58.

———. "Biolinguistics and the Human Capacity." Chomsky.info (2004), www.chomsky.info/talks/20040517.htm (accessed February 8, 2008).

———. "Of Minds and Language." *Biolinguistics* 1 (2007), www.biolinguistics .eu/index.php/biolinguistics/article/view/19/21%2021.

Christiansen, M. H. "Brains, Genes, and Language Evolution." Evolang Conference, Utrecht, Netherlands, April 14–17, 2010.

Christiansen, M. H. and N. Chater. "Language as Shaped by the Brain." *Behavioral and Brain Sciences* 31 (2008): 489–509.

Conway, C. M. and M. H. Christiansen. "Sequential Learning in Non-Human Primates." *Trends in Cognitive Sciences* 5 (2001): 539–546.

Corballis, M. C. "Mental Time Travel and the Shaping of Language." *Experimental Brain Research* 192 (2009): 1491–1499.

Crompton, R. H. and Thorpe, S. K. "Response to Comment on 'Origin of Human Bipedalism as an Adaptation for Locomotion on Flexible Branches.'" *Science* 318 (2007).

Darwin, C. *On the Origin of Species*. Many editions, 1859.

Dawkins, R. *The Selfish Gene*. Oxford: Oxford University Press, 1976.

Dawkins, R. and J. B. Krebs. "Animal Signals: Information or Manipulation?" *Behavioral Ecology: An Evolutionary Approach*. Edited by J. Krebs and N. Davies. Malden, MA: Blackwell Scientific Publications, 1978: 282–309.

de Boer, B. "Air Sacs and Speech." Evolang Conference, Barcelona, Spain, March 12–15, 2008.

———. "Modelling Vocal Anatomy's Significant Effect on Speech." *Journal of Evolutionary Psychology* 8 (2010): 351–366.

de l'Etang, A. M.; P. J. Bancel; and J. D. Bengtson. "The Proto-Sapien Words PAPA, MAMA and KAKA and the Origin of Kinship Systems." Cradle of Language Conference, Stellenbosch, South Africa, November 6–10, 2006.

de Waal, F. *Chimpanzee Politics: Power and Sex Among Apes*. New York: Unwin, 1982.

Deacon, T. *The Symbolic Species: The Co-Evolution of Language and Brain.* New York: W.W. Norton, 1997.

————. Ways to Protolanguage Conference, Torun, Poland, September 21–23, 2009

————. Relaxed Selection, Complexity and Language. *Darwinism Inside-Out.* Elizabethtown presentation, 2010.

Dedieu, D. and R. Ladd. "Linguistic Tone Is Related to the Population Frequency of the Adaptation Haplogroups of Two Brain Size Genes, *ASPM* and *Microencephalin.*" *PNAS* 104 (2007), 10944–1049.

Dessalles, J. L. "Language and Hominid Politics." *The Evolutionary Emergence of Language: Social Function and the Origin of Linguistic Form.* Edited by C. Knight and J. R. Hurford. Cambridge, U.K.: Cambridge University Press, 2000: 62–82.

————. *Why We Talk.* New York: Oxford University Press, 2007.

Dunbar, R. *Grooming, Gossip and the Evolution of Language.* Cambridge, MA: Harvard University Press, 1997.

Eizirik, E. M. "Molecular Phylogeny and Dating of Early Primate Divergence. *Anthropoid Origins.* Edited by C. Ross and R. Kay. New York: New Visions Kluwer Academic Press/Plenum Publishers, 2004.

Endress, A. D.; D. Cahill; S. Block; J. Watumull; and M. D. Hauser. "Evidence of an Evolutionary Precursor to Human Language Affixation in a Non-Human Primate." *Biology Letters* 5 (2009): 749–751.

Falk, D. *Finding Our Tongues: Mothers, Infants, and the Origins of Language.* New York: Basic Books, 2009.

Feher, O.; H. Wang; S. Saar; et al. (2009). "*De novo* Establishment of Wild-Type Song Culture in the Zebra Finch." *Nature*, 564-568.

Fitch, W. Tecumseh. "Prolegomena to a Future Science of Linguistics." *Biolinguistics* 3 (2010), www.biolinguistics.eu/index.php/biolinguistics/article/view/133/132.

————. *The Evolution of Language.* Cambridge: Cambridge University Press, 2010.

Fitch, W. Tecumseh; Marc Hauser; and Noam Chomsky. "The Evolution of the Language Faculty: Clarifications and Implications." *Cognition* 97 (2005): 179–210.

Fitch, W. Tecumseh; L. Huber; and T. Bugnyar. "Social Cognition and the Evolution of Language: Constructing Cognitive Phylogenies." *Neuron* 65 (2010): 795–814.

Fletecher, J.; M. Zwick; M. Doebeli; and D. Wilson. "What's Wrong with Inclusive Fitness." *Trends in Ecology & Evolution* 21 (2006): 597–598.

Fodor, J. and M. Piatelli-Palmarini. *What Darwin Got Wrong.* New York: Farrar, Straus and Giroux, 2010.

Freese, J. "The Problem of Predictive Promiscuity in Deductive Applications of Evolutionary Reasoning to Intergenerational Transfers: Three Cautionary Tales." Symposium on Family Issues, Penn State, October 2006.

Gardenfors, P. *How Homo Became Sapiens: On the Evolution of Thinking.* New York: Oxford Universiity Press, 2004.

Glazko, G. and M. Nei. "Estimation of Divergence Times for Major Lineages of Primate Species." *Molecular Biological Evolution* 20 (2003): 424–434

Gold, R. and M. Faust. "Right Hemisphere Dysfunction and Metaphor Comprehension in Young Adults with Asperger Syndrome." *Journal of Autism Developmental Disorder* 40 (2010): 800–811.

Gold, R.; M. Faust; and A. Goldstein. "Semantic Integration During Metaphor Comprehension in Asperger Syndrome." *Brain and Language* 113 (2010): 124–34.

Goldin-Meadow, Susan. "Gesture Adds More Than Structure." Evolang Conference, Barcelona, Spain, March 12–15, 2008.

Gong, T.; J. Minett; and W. S. Wang. "A Framework Triggering Displacement in Human Language." Ways to Protolanguage Conference, Torun, Poland, September 21–23, 2009.

Goodnight, C. "Multilevel Selection: The Evolution of Cooperation in Non-Kin Groups." *Population Ecology* 47 (2005): 3–12.

Goren-Inbar, N.; N. Alperson; M. Kislev; et al. "Evidence of Hominin Control of Fire at Gesher Benot Ya'aqov." *Science* 30 (2004): 725–727.

Graber, R. R. and W. W. Cochran. "An Audio Technique for the Study of the Nocturnal Migration of Birds." *Wislon Bulletin* 71 (1959): 220–236.

Green, R. E.; A. S. Malaspinas; J. Krausse; et al. "A Complete Neandertal Mitochondrial Genome Sequence Determined by High-Throughput Sequencing." *Cell* 134 (2008): 416–426.

Grossman, T. and M. H. Johson. "Selective Prefrontal Cortex Responses to Joint Attention in Early Infancy." *Biology Letters Neurobiology* 6 (2010): 540–543.

Haidt, J. "The New Synthesis in Moral Psychology." *Science* 18 (2007): 998–1002.

Hamilton, W. "The Evolution of Altruistic Behavior." *American Naturalist* 97 (1963): 354–356.

———. "The Genetic Evolution of Social Behavior." *Journal of Theoretical Biology* 7 (1964): 1–52.

Hauser, Marc D.; Noam Chomsky; and W. Tecumseh Fitch. "The Language Faculty: What Is It, Why Has It, and How Did It Evolve?" *Science* (2002): 1569–1579.

Hughes, J. F.; H. Skaletsky; T. Pyntikova; et al. "Chimpanzee and Human Y Chromosomes Are Remarkably Different in Structure and Gene Content." *Nature* 463 (2010): 463–539.

Hurford, J. R. "The Origin of Noun Phrases: Reference, Truth, and Communication." *Lingua* 117 (2007): 527–542.

———. "Pragmatics, Storage & Computation in the Evolution of Syntax." Ways to Protolanguage Conference, Torun, Poland, September 21–23, 2009.

Isler, K. and C. Van Schaik. "Why Are There So Few Smart Mammals (But So Many Smart Birds)?" *Biology Letters* 5 (2009): 125–129.

———. "The Expensive Brain: A Framework for Explaining Evolutionary Changes in Brain Size." *Journal of Human Evolution* 57 (2009): 392–400.

Jablonka, E. and M. J. Lamb. *Evolution in Four Dimensions: Genetic, Epigenetic, Behavioral, and Symbolic Variation in the History of Life.* Cambridge: MIT Press, 2005.

Jackendoff, Ray. *Foundations of Language.* Oxford: Oxford University Press, 2002.

James, Henry. *The Wings of the Dove.* Various editions, 1902.

Karpf, A. *The Human Voice: How This Extraordinary Instrument Reveals Essential Clues About Who We Are.* New York: Bloomsbury USA, 2006.

Kazakov, D. and M. Bartlett. "Could Navigation Be the Key to Language." University of York, www-users.cs.york.ac.uk/~kazakov/papers/EELC-05.pdf (accessed July 13, 2010).

Kegl, J. "The Whole Shebang: Looking at Language as a Human Trait." Cradle of Language Conference, Stellenbosch, South Africa, November 6–10, 2006.

Kegl, J.; A. Senghas; and M. Coppola. "Creation Through Contact: Sign Language Emergence and Sign Language Change in Nicaragua." *Language Creation and Change.* Edited by M. DeGraff. Cambridge: MIT Press, 1999.

Kenneally, C. *The First Word: The Search for the Origins of Language.* New York: Viking, 2007.

Kevinropelato. "Rock Throwing Chimp." YouTube, www.youtube.com/watch?v=fDkudPGqTK8 (accessed July 27, 2010).

Kirby, S.; H. Cornish; and K. Smith. "Cumulative Cultural Evolution in the Laboratory: An Experimental Approach to the Origins of Structure in Human Language." *PNAS* 105 (2008): 10681-10686.

Knight, C. "Language and Revolutionary Consciousness." *The Transition to Language.* Edited by A. Wray. Oxford: Oxford University Press, 2002: 138–160.

Konopka, G. and J. M. Bomar. "Human-Specific Transcriptional Regulation of CNS Development Genes by FOXP2." *Nature* 462 (2009): 213–217.

Kuwenhoven, A. P. "World's Oldest Spears." *Archaeology* 50 (1997).

Lakoff, G. and M. Johnson. *Metaphors We Live By.* Chicago: University of Chicago Press, 1980.

Lamarck, J. B. *Philosophie Zoologique*. Translated by Hugh Elliot. Chicago: University of Chicago Press, 1809/1984.

Leakey, M. "Pliocene Footprints in the Laetoli Beds at Laetoli, Northern Tanzania." *Nature* 278 (1979): 317–323.

Leavens, D. A.; W. D. Hopkins; and K. A. Bard. "Understanding the Point of Chimpanzee Pointing: Epigenesis and Ecological Validity." *Current Directions in Psychological Science* 14 (2005): 185–189.

Lenneberg, E. H. *Biological Foundations of Language*. New York: Wiley, 1967.

Levin, N. E.; S. W. Simpson; S. Semaw; and M. Rogers. "Isotopic Evidence for Plio-Pleistocene Environmental Change at Gona, Ethiopia." *Earth Science Letters* 219 (2004): 93–110.

Levitin, D. *This Is Your Brain on Music*. New York: Dutton, 2006.

Lewin, R. and G. H. Curtis. *Java Man: How Two Geologists' Dramatic Discoveries Changed Our Understanding of the Evolutionary Path to Modern Humans*. New York: Scriber, 2000.

Lieberman, E.; J. B. Michel; J. Jackson; T. Tang; and M. A. Nowak. "Quantifying the Evolutionary Dynamics of Language." *Nature* 449 (2007): 713–716.

Locke, J. L. and B. Bogin. "Language and Life History: A New Perspective on the Development and Evolution of Human Language." *Behavioral & Brain Sciences* 29 (2006): 259–280.

Lovejoy, C.; B. Latimer; G. Suwa; et al. "Combining Prehension and Propulsion: The Foot of Ardipithecus Ramidus." *Science* 326 (2009): 72.

Maasch, K. A. and B. Saltzman. "A Low-Order Dynamical Model of Global Climactic Variability Over the Full Pleistocene." *Journal of Geophysical Research* 95 (1990): 1955–1963.

Marchetti, G. "A Presentation of Attentional Semantics." *Cognitive Processing* 7 (2006): 163–194.

Marcus, G. F. *Kluge: The Haphazard Evolution of the Human Mind*. Boston: Houghton Mifflin Co., 2008.

Margoliash, D. and H. C. Nusbaum. "Language: The Perspective from Organismal Biology." *Trends in Cognitive Science* 13 (2009): 505–510.

Masataka, N. "The Origins of Language and the Evolution of Music: A Comparative Perspective." *Physics of Life Reviews* 6 (2009): 11–22.

Maynard-Smith, J. "Game Theory and the Evolution of Behaviour." *Proceedings of the Royal Society of London: Biological Sciences* 205 (1979): 475–488.

———. *Evolution and the Theory of Games*. Cambridge: Cambridge University Press, 1982.

McBrearty, S. and A. Brooks. "The Revolution that Wasn't." *Journal of Human Evolution* 39 (2000): 453–563.

Mundy, P. and L. Newell. "Attention, Joint Attention, and Social Cognition."
    Current Directions in Psychological Science 16 (2007): 269–274.

Nevell, L. and B. Wood. "Cranial Base Evolution within the Hominin Clade."
    Journal of Anatomy 212 (2008): 455–468.

Nobel, J.; J. P. de Ruiter; and K. Arnold. "From Monkey Alarm Calls to Human
    Language: How Simulations Can Fill the Gap." *Adaptive Behavior* 18
    (2010): 66–82.

Nova, "Dogs Decoded." Television program first broadcast November 9, 2010.
    Produced and directed by Dan Child.

Nunney, L. "Lineage Selection: Natural Selection for the Long-Term Benefit."
    Levels of Selection in Evolution. Edited by L. Keller. Princeton, NJ: Princeton
    University Press, 1999: 238–252

Oller, D. K. and U. Griebel. "How the Language Capacity Was Naturally Selected:
    Altriciality and the Long Immaturity." *Behavioral and Brain Sciences* 29
    (2005): 293–294.

Ostler, N. *Empires of the Word: A Language History of the World*. New York:
    HarperCollins, 2005.

Oudeyer, P. Y. and F. Kaplan. "Language Evolution as a Darwinian Process:
    Computational Studies." *Cognitive Process* 8 (2010): 21–35.

Page, S. and M. Goodman. "Catarrhine Phylogeny Noncoding DNA Evidence
    for a Diphyletic Origin of the Managabeys and for a Human-Chimpanzee
    Clade." *Molecular Phylogenetics & Evolution* 18 (2001): 14–25.

Penny, D. "The Continuity of Mind from Great Apes to Humans." *New Zealand
    Science Review* 67 (2010): 63–69.

Piattelli-Palmarini, M. "Novel Tools at the Service of Old Ideas." *Biolinguistics* 2
    (2008), http://biolinguistics.eu/index.php/biolinguistics/article/view/61/70.

Pika, S. "Gestures of Apes and Pre-Linguistic Human Children: Similar or
    Different?" *First Language* 28 (2008): 116–140.

Pinker, S. *The Language Instinct: How the Mind Creates Language*. New York:
    William Morrow, 1994.

———. *The Stuff of Thought: Language as a Window into Human Nature*. New
    York: Viking, 2007.

Pinker, S. and P. Bloom. "Natural Language and Natural Selection." *Behavioral
    and Brain Sciences* 13 (1990): 707–784.

Pontzer, H.; C. Rolian; G. P. Rightmire; et al. "Locomotor Anatomy and
    Biomechanics of the Dmanisi Hominins." *Journal of Human Evolution* 58
    (2010): 492–504.

Posner, M. *The Foundations of Cognitive Science*. Cambridge, MA: MIT Press,
    1989.

Progovac, L. "If Syntax Evolved, What Might Be Some Vestiges of Its Evolution in Modern-Day Languages?" *The Emergence of Language in the Child and in the Species.* New York, 2007.

Puglisi, A.; A. Baronchelli; and V. Loreto. "Cultural Route to the Emergence of Linguistic Categories." *PNAS* 105 (2008): 7936–7940.

Pulvermuller, F. "Brain-Language Research: Where Is the Progress?" *Biolinguistics* 4 (2010): 255–288.

Pulvermuller, F. and R. Assadollahi. "Grammar or Serial Order?: Discrete Combinatorial Brain Mechanisms Reflected by the Syntactic Mismatch Negativity." *Journal of Cognitive Neuroscience* 19 (2007): 971–980.

Reboul, A. "Cooperation and Competition in Apes and Humans: A Comparative and Pragmatic Approach to Human Uniqueness." *Pragmatics & Cognition* 18 (2010): 422–441.

Richardson, R. D. *William James: In the Maelstrom of American Modernism.* Boston: Houghton Mifflin Co., 2006.

Riede, T.; J. H. Fisher; and F. Goller. "Sexual Dimorphism of the Zebra Finch Syrinx Indicates Adaptation for High Fundamental Frequencies in Males." PLoS ONE, www.plosone.org/article/info%3Adoi%2F10.1371%2Fjournal .pone.0011368 (accessed November 18, 2010).

Ross, D. "*H. Sapiens* as Ecologically Special; What Does Language Contribute?" *Language Sciences* 29 (2007): 710–731.

Ross, E. D. "Cerebral Localization of Functions and the Neurology of Language: Fact vs. Fiction or Is It Something Else?" *The Neuroscientist* 16 (2010): 1–22.

Rult, C. J. "The Mouse Genome Database (MGD): Mouse Biology and Model Systems." *Nucleic Acids Research* 36 (2007): D724–D728.

Rundblad, G. and D. Annaz. "The Atypical Development of Metaphor and Metonymy in Children with Autism." *Autism* 14 (2010): 29–46.

Sample, I. "Chimp Who Threw Stones at Visitors Showed Human Traits, Says Scientist." *The Guardian*, March 9, 2009.

Satta, Y. et al. "Ancestral Population Sizes and Species Divergence Times in the Primate Lineage on the Basis of Intron and BAC End Sequences." *Journal of Molecular Evolution* 59 (2010): 478–487.

Sawyer, G. and V. Deak. *The Last Human: A Guide to Twenty Two Species of Extinct Humans.* New Haven: Yale University Press, 2007.

Seyfarth, R. M. and D. L. Cheney. "Primate Vocal Communication." *Primate Neuroethology.* Edited by M. L. Platt and A. A. Ghazanfar. New York: Oxford University Press, 2010: 84–97.

Sigman, M. and C. Kasari. "Joint Attention Across Contexts in Normal and Autistic Children." *Joint Attention: Its Origins and Role in Development.* Edited by C. Moore and P. Dunham. Hillsdale, NJ: Lawrence Erlbaum, 1995.

Skinner, B. F. *Verbal Behavior*. New York: Appleton-Century-Crofts, 1957.

Sperber, D. and D. Wilson. *Relevance: Communication and Cognition*. Oxford: Blackwell, 1986.

Stauffer, R. et al. "Human and Ape Molecular Clocks and Constraints on Paleontological Hypotheses." *Journal of Heredity* 92 (2001): 469–474.

Steiper, M. et al. "Genetic Data Support the Hominoid Slowdown and Early Oligocene Estimate for the Hominoid-Cercopithecoid Divergence." *Proceedings of the National Academy of Sciences* 101 (2004): 17021–17026.

Stout, D.; S. Semaw; M. J. Rodgers; and D. Cauche. "Technological Variation in the Earliest Oldowan from Gona, Afar, Ethiopia." *Journal of Human Evolution* 58 (2010): 474–491.

Stout, D.; N. Toth; K. Schick; and T. Chaminade. "Neural Correlates of Early Stone Age Toolmaking: Technology, Language and Cognition in Human Evolution." *Phil Trans R Soc Lond B Biol Sci* 363 (2008): 1939–49.

Stromswold, K. Presentation at conference, Santo Domingo, Dominican Republic, February 23–25, 2007.

Thorpe, S. K.; R. H. Crompton; et al. "Dimensions and Moment Arms of the Hind- and Forelimb Muscles of Common Chimpanzees." *American Journal of Physical Anthropology* 110 (1999): 179–199.

Tomasello, M. *Origins of Human Communication*. Cambridge: MIT Press, 2008.

Tomasello, M.; M. Carpenter; and U. Liszkowski. "A New Look at Infant Pointing." *Child Development* 78 (2007): 705–722.

Trauth, M. H.; M. A. Maslin; A. Deino; et al. "High- and Low-Latitude Forcing of Plio-Pliestocene East African Climate and Human Evolution." *Journal of Human Evolution* 53 (2007): 475–486.

Tylen, K.; E. Weed; M. Wallentin; A. Roepstorff; and C. D. Frith. "Language as a Tool for Interacting Minds." *Mind and Language* 25 (2010): 3–29.

Wade, Nicholas. "In Lice, Clues to Human Origin and Attire." *The New York Times*, March 8, 2008.

———. "Clues of Britain's First Humans." *The New York Times*, July 7, 2010.

Wallace, A. "On the Tendency of Varieties to Depart Indefinitely from the Original Type." *Galileo's Commandment: An Anthology of Great Science Writing*. Edited by E.B. Bolles. W. H. Freeman and Company, 1997: 389–399.

Walrath, D. "Gender, Genes, and the Evolution of Human Birth." *Feminist Anthropology: Past Present and Future*. Edited by P. L. Geller and M. K. Stockett. Philadelphia: University of Pennsylvania Press, 2006: 55–71.

Warren, W.C. et al. "Genome Analysis of the Platypus Reveals Unique Signature of Evolution." *Nature* 453 (2008): 175–183.

Weaver, W. and C. E. Shannon. *The Mathematical Theory of Communication*. Urbana: University of Illinois Press, 1949.

Weinstock, G. M.; G. E. Robinson; et al. "Insights in Social Insects from the Genome of the Honeybee Apis Meliferra." *Nature* 443 (2006): 931–949.

West, S. A.; C. El Mouden; and A. Gardner. "Sixteen Common Misconceptions About the Evolution of Cooperation in Humans." *Evolution and Human Behavior* (2010), doi:10.1016/j.evolhumbehav. 2010.08.001.

Whitehead, C. "The Culture Ready Brain." *Social Cognitive and Affective Neuroscience* 5 (2010): 168–179.

Whiten, A. "Imitation of the Sequential Structure of Actions by Chimpanzees (Pan Troglodytes)." *Journal of Comparative Psychology* 112 (1998): 3–14.

Wildeman, D. et al. "Implications of Natural Selection in Shaping 99.4%, Nonsynonymous DNA Identity Between Humans and Chimpanzees; Enlarging Genus *Homo*." *Proceedings of the National Academy of Sciences* 100 (2003): 7181–7188.

Wilkinson, R. D.; M. E. Steiper; C. Soligo; et al. "Dating Primate Divergences Through an Integrated Analysis of Paleontological and Molecular Data." *Systems Biology* 60 (2011): 16–31.

Wilson, D. S. *Darwin for Everyone.* New York: Delacourt Press, 2007.

Wilson, D. S. and E. O. Wilson. "Rethinking the Theoretical Foundation of Sociobiology." *The Quarterly Review of Biology* 82 (2007): 327–348.

Wilson, E. O. *Sociobiology: The New Synthesis.* Cambridge: Harvard University Press, 1975.

Wilson, E. O. and B. Hölldobler. "Eusociality: Origin and Consequences." *Proceedings of the National Academy of Sciences of the United States of America* 102 (2005): 13367–13371.

Wrangham, R. and R. Carmody. "Human Adaptation to the Control of Fire." *Evolutionary Anthropology* 19 (2010): 187–199.

Wray, G. A. "The Evolutionary Significance of *Cis*-Regulatory Mutations." *Nature Reviews Genetics* 8 (2007): 206–216.

Wynn, J. "Influence of Plio-Pleistocene Aridification on Human Evolution: Evidence from Paleosols of the Turkana Basin, Kenya." *American Journal of Physical Anthropology* 123 (2004): 106–118.

Young, R. W. "Evolution of the Human Hand: The Role of Throwing and Clubbing." *Journal of Anatomy* 202 (2003): 165–174.

Zahavi, A. and A. Zahavi. *The Handicap Principle.* Oxford: Oxford University Press, 1997.

Zuidema, W. and B. de Boer. "The Evolution of Combinatorial Phonology." *Journal of Phonetics* 37 (2009): 125–144.

Zuidema, W. and A. Verhagen. "What Are the Unique Design Features of Language? Formal Tools for Comparative Claims." *Adaptive Behavior* 18 (2010): 48–65.